Building Design
Cost Manageme

IN
T

Building Design Cost Management

David Jaggar
Emeritus Professor of Construction Economics and former Deputy Director,
School of Built Environment, Liverpool John Moores University

Andy Ross
Head of the construction section, Liverpool John Moores University

Jim Smith
Senior Lecturer in Property and Construction Economics, Faculty of Architecture,
Building and Planning, The University of Melbourne

and

Peter Love
School of Management Information Systems, Edith Cowan University (Australia)

Blackwell
Science

© 2002 by Blackwell Science Ltd,
a Blackwell Publishing Company
Editorial Offices:
Osney Mead, Oxford OX2 0EL, UK
    Tel: +44 (0)1865 206206
Blackwell Science, Inc., 350 Main Street,
Malden, MA 02148-5018, USA
    Tel: +1 781 388 8250
Iowa State Press, a Blackwell Publishing
Company, 2121 State Avenue, Ames,
Iowa 50014-8300, USA
    Tel: +1 515 292 0140
Blackwell Publishing Asia Pty Ltd,
550 Swanston Street, Carlton South, Melbourne,
Victoria 3053, Australia
    Tel: +61 (0)3 9347 0300
Blackwell Wissenschafts Verlag, Kurfürstendamm
57, 10707 Berlin, Germany
    Tel: +49 (0)30 32 79 060

The right of the Author to be identified as the
Author of this Work has been asserted in
accordance with the Copyright, Designs and
Patents Act 1988.

All rights reserved. No part of this publication
may be reproduced, stored in a retrieval system,
or transmitted, in any form or by any means,
electronic, mechanical, photocopying, recording
or otherwise, except as permitted by the UK
Copyright, Designs and Patents Act 1988,
without the prior permission of the publisher.

First published by Blackwell Science, 2002

Library of Congress
Cataloging-in-Publication Data
is available

ISBN 0-632-05805-6

A catalogue record for this title is available from
the British Library

Produced and typeset by Gray Publishing,
Tunbridge Wells, Kent
Printed and bound in Great Britain by
MPG Books Ltd, Bodmin, Cornwall

For further information on
Blackwell Science, visit our website:
www.blackwell-science.com

LIBRARY

REGIONAL
TE
CO

class no: 692·5 JAG

acc. no: 25181

Contents

It is better to be approximately correct than to be precisely wrong. Accuracy is preferred to precision.

(Hicks 1992)

Foreword

It may be said – with about as much exaggeration as may be normally attributed to such statements, but no more – that the publication, 50 years ago, of the Ministry of Education Building Bulletin No. 4 introduced the concept of elemental cost planning to the UK construction industry.

The preface to the bulletin says that with the need for a great number of new schools, it is 'essential to explore every possible means of reducing the cost, increasing the speed of school building whilst maintaining standards of quality and educational efficiency'. This is as true today as it was then.

The bulletin described a 'new approach towards a costing technique for use by architects and surveyors'. The fact that the quantity surveying profession made the technique their own may explain the changing influence of the two professions over the past half-century.

The key to this innovation was the concept of elements. Again, it is worth quoting from the bulletin:

> An architect tends to think in terms of functions and of the means by which he can perform those functions. For example, he sees as one function the exclusion of rain and weather and he looks to a roof to perform this task. It is, for the purposes of cost analysis, immaterial to him whether the roof be of timber and tiles or of concrete and asphalt. He is primarily concerned to know how much it has cost to roof in the building.
>
> Thus, where the estimator builds up his tender by adding together a large number of relatively small items, classified by trades, cost analysis must reverse this process and break down the tender into groups of material and labour classified according to the functions they perform. These groups have been described as elements.

Thus an element is defined as a part of a building that fulfils a specific function or functions irrespective of its design, specification or construction.

The cost planning technique has served the profession well in offering value-added services to clients and the concept of elements has been incorporated into the development of life cycle costing and value management. Furthermore the current changes to procurement practice are making the cost plan central to the entire process.

On Private Finance Initiative schemes, the provider's ability to control costs through the life of the building is central to the profitability of the contract. Cost planning, value engineering and life cycle costing are essential tools in the delivery of optimum solutions. All these techniques are based on the use of clearly defined functional elements.

The existence of a well-founded cost plan is the key to the delivery and control of Partnering and Prime contracting projects, indeed it will be the only method available to set the target costs and administer the contract.

The Building Cost Information Service (BCIS) was set up by the Royal Institution of Chartered Surveyors in 1961 to provide the cost information in an elemental form so that the technique of elemental cost planning could be developed by quantity surveying firms. BCIS has been the keeper of the industry's database and the guardian of elemental orthodoxy ever since.

I was therefore delighted for BCIS to be involved with this new publication with its novel approach to teaching the subject by the use of the accompanying website to allow students to manipulate data alongside the instructional text.

I am sure that the use of the website will make the text more accessible to students and as 'doing is the key to learning' it will help them with their understanding of the techniques and concepts described.

BCIS now has more customers accessing information online than in hardcopy so we were happy to allow information from the BCIS *Online* service to be used in the website that accompanies this book.

The authors are to be congratulated on this timely and ground-breaking work that takes the teaching of an important subject into the Internet age.

Joe Martin DMS FRICS MIMgt
Executive Director
Building Cost Information Service Ltd.

Preface

Unpredictability can result from the client's bureaucratic system of management. Under this management system, many levels within the hierarchy of the organisation can be assigned the responsibility for deciding the range of services required: in-house professionals, financial officers, and different user groups. At the chief executive level, family members and friends sometimes get into the act. The difficulties are intensified when the building problem is complex or when the project is for a relatively new building type such as many organisations are now constructing in order to capitalise on advances in building technology …

Gutman (1988)

Some people consider that there is nothing more to learn about cost planning because there are lots of books written on this important topic. The authors agree that there are many excellent texts in building cost planning, but we believe there is still a need to broaden the focused cost control skills of the cost planner to appreciate and embrace client, community and design team interactions. To neglect this aspect is to practice cost planning in one dimension—an approach that is totally unacceptable in today's client change driven world. This can be considered the second dimension to modern cost planning. The third dimension is to significantly integrate modern web-based construction cost and other databases into mainstream cost planning practice.

Our approach aims to add these two additional dimensions to traditional building cost planning to transform it into building design cost management as noted in the title of this book.

So this book has been written as an aid, to both students and practitioners, who have an interest in both understanding and being able to apply the techniques of design cost control as developed since the early 1950s. These methods are of particular importance in that, if applied correctly, they can ensure better value through more effective cost and resource management during the most influential phase of the construction process, the design stage. Clearly this stage offers maximum opportunities to produce both effective and efficient building solutions to all those concerned with both design and construction. As each solution is put in place during the maturation of design and construction, then alternatives become, at worst, closed down, or, at best, expensive and time consuming to change. Put very simply, at the inception of a building project, every solution in the world is available for consideration, whereas once construction is completed, no further solutions are open, without some form of remedial action being taken.

In fact, it is useful for the design team to consider that a series of pre-briefing events will have occurred within the client's organisation before the building design and cost planning stages have commenced. The pre-design or pre-briefing stages are crucial steps in the life of a project and will determine whether a built solution will be the *solution* to the client's problems. Without analysing these steps in detail it is informative to review the pre-briefing steps that resulted in the *decision to build*. These steps are illustrated in Fig. 0.1. Needless to say, if a strategic decision is made to adopt an organisational solution, then a building project will not commence. Our work in the design team will only commence if a decision to follow a property solution is made. However, as noted in Fig. 0.1, the built solution is likely to also involve a related organisational change.

Fact file

At the beginning of the nineteenth century, when Britain was at war with Napoleon, there was a need to house the large British army, on the South coast, in permanent barracks. The existing system of procuring building works at that time was to negotiate separately with each individual master craftsman carrying out the various trades and price was not generally agreed before the start of construction. Apart from the obvious financial uncertainty of such an approach, the major drawback was the time consuming nature of the process, together with the inconvenient multi-point, client–craftsmen lines of communications. As a result, the idea of main general contracting was born establishing there was a single point of contact, with one main or general contractor. It was also quickly realised that such an approach would facilitate the determination of the price before work actually commenced and that financial competition could be achieved by inviting a number of such contractors to submit their tender bids for carrying out the works in advance. A further advantage was that the successful contractor would be responsible for carrying much of the risk. For many clients the avoidance of risk is a very attractive benefit, although as will be expanded, this may not always be as beneficial as it might at first seem. Clearly, this approach was a huge success and transformed the whole approach to the procurement of buildings, and remains the basis of procurement in the UK, and in many other countries of the world.

This is not to say that those concerned with the actual building on site are not responsible for contributing to the efficient and effective construction of the building on which they are working. In fact, it is still essential to maintain the initial budget during the construction stage. Therefore, it is critical that we have a construction stage cost control process, which achieves maximum resource utilisation in terms of materials, plant, labour and time. The procurement methods adopted by the construction industry assist in this process—by the use of financial competition to discourage such organisations from not being as efficient as they might otherwise be. However, to emphasise the point, if the design is inherently inefficient, then such on-site efficiencies will only achieve marginal benefits, at best.

It is also important to stress that the process of achieving efficient and effective design and construction solutions is not just about creating the minimum physical solution, in terms of bricks and mortar, which at first glance might seem the optimal solution, but rather by being aware of the dynamics of the construction process itself. To illustrate the point, a minimum *in-situ* concrete floor slab specification in a multi-storey office block may need less concrete, but will probably cost more, ultimately, as supporting formwork may have to be left in at the various floors in order to support the dead loads of the concrete slabs being cast above.

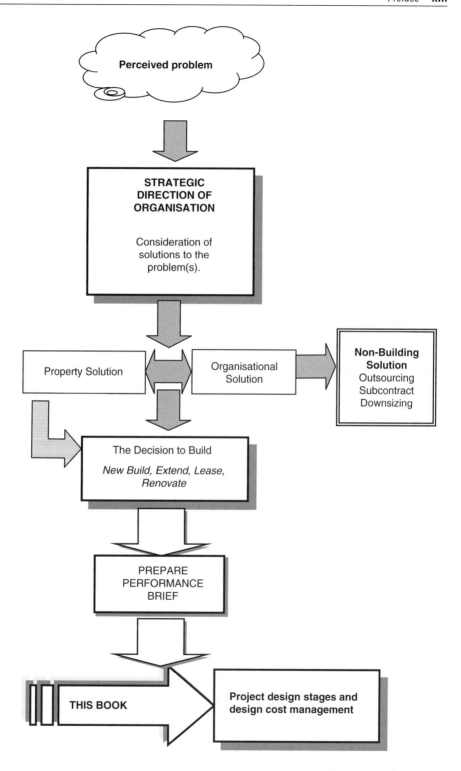

Figure 0.1 Strategic pre-briefing steps.

Clearly this restricts the slab formwork and supports to one use, which has significant cost and time implications. So buildability, the understanding of the construction process and how it impacts on resource utilisation, is of paramount importance to any individual or organisation concerned with design and cost control. In fact, this is partly the reason for the increase in the use of Design and Build and Management style methods of procurement, compared with the more traditional and historic approach, based on the separation of design and construction, both in terms of chronology and responsibility. To be blunt, this is partly because of the failure of quantity surveyors, concerned with giving financial advice during the early design stages, not being fully aware of the buildability issues and how they affect on-site construction.

It is also worth stressing that, whatever design and construction solution is selected, it will be a compromise, as the various values being sought will often conflict with each other. For example, a hard wearing floor finish may not be aesthetically pleasing, so it is important that design cost management knows its place, success being not necessarily finding the least cost solution, but rather that which provides the best solution for those who will ultimately use it.

For example, the design cost management proposed by the Chief Quantity Surveyor at a local authority, where one of the authors spent his formative years is salutary. His advice to the architects was, 'You can have any shape you like as long as it's square and has a flat roof' was hardly going to lead to the establishment of buildings embracing the opportunities and values that needed to be explored by the design team.

There are also exceptions where the apparent lack of effective design cost management has ultimately facilitated the production of highly successful buildings, which in terms of financial management have been disastrous. The Sydney Opera House is perhaps the best example where, had an effective system of design cost control been put in place, such a spectacular solution would never have been achieved. For the record, the final cost was 35 times that of the original estimate, hardly effective design cost management. And, just to show there is nothing new, it needed the assistance of a State lottery to bring it to a successful conclusion!

However, there are many examples where such successful outcomes have not been achieved, which can be confirmed by perusal of the technical press. A typical example is the new British Library, which was predicted to cost £164 million in 1978 but ultimately cost in the order of 300% more on completion in 1997 (*Construction News* 1997).

Design cost management is essential and necessary if building clients are to be confident that a budget is reliable, will be adhered to, and that the ultimate solution achieves the best set of values possible for that client. In turn, the design team must deliver a cost control process that achieves its aim—a final building within cost, on time and of the appropriate quality. That is, the timeless *project objectives*, as shown in Fig. 0.2.

We aim to inform the reader of the nature of design cost management and how it works in a modern client-based technological environment. It takes the reader through its historical evolution in order to explain why it has developed in the way it has and discusses how current innovations have impacted on the modern process of design cost management, some more successfully than others. References are made to developments in the application of Information Technology on

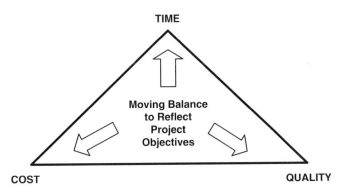

Figure 0.2 Project objectives (Rowlinson and McDermott 1999).

design cost management, and in particular the development and application of cost modelling. Consideration is also given to the application of techniques designed to improve the objectivity of the design cost management process, such as value and risk management.

Important issues such as of the changing methods of procurement are discussed. Interestingly, these new and possibly novel procurement methods are aimed at creating greater understanding, transparency and harmony between the various organisations involved in the design and construction of buildings. These innovations were advocated by Latham (1994) and more recently by Egan (1998). We will be focusing on the role of design cost management in these changing environments. For balance, the limitations of the design cost control process are also considered and some suggestions made as to how such limitations might be overcome.

A central focus of the text is the demonstration of the application of the process of design cost control to a real building, where the various issues are discussed, and solutions developed. This allows the reader to understand and reinforce, through example, how design cost control works in a simulated project environment. The reader will be taken through an actual building project where the various concepts are considered, solutions developed and design cost control information produced, for use by the various participants involved in the design and construction process.

Fact file

A unique feature, that has been developed to support the text, is an accompanying dedicated website which provides the reader with a dynamic tool which enables him or her to carry out the design cost control process. The website has been developed in conjunction with the national Building Cost Information Service which, as part of the Royal Institution of Chartered Surveyors, acts as custodian of financial and economic information of relevance to the building industry, and equally importantly, researches and develops management techniques of buildings and their cost implications, with considerations ranging from global, to those of specific buildings. The website allows the user to find, home in on and access financial information of relevance to the actual building project referred to in the text. The reader can then develop his or her cost description of the project and compare it with the suggested solution developed by the authors. The website address is www.bdcm.co.uk and is structured to contain briefing and design information about the case study which was a project built recently in Liverpool, UK. This information is available in a format that would be evidenced in practice and develops in detail as the reader progresses through the text. The website also contains cost information and cost models that allow the reader to test the authors' assumptions but also develop cost plans and check of their own. Increasingly, we will call upon you to refer to the webpage for guidance and an updated version of the printed examples contained in this book. When you see the following symbol in the text, we would like you to go to the webpage and follow the advice given in that particular section:

The website also has a number of interesting links and slides that lecturers can use to support the teaching of Building Design Cost Management. A list of the information that is contained on the site is given in Appendix 1.

The use of a dedicated webpage is an exciting development that has a number of benefits to the practitioner and student of Building Design Cost Management as follows:

- It enables an understanding of the benefits of web-supported information technology as part of the developing e-commerce culture generally and in design cost management specifically.
- It provides a dynamic modelling facility, allowing the application of rapid and easily accomplished iterations as a necessary process in seeking optimal solutions in complex problem solving, such as design cost control.
- It encourages a better appreciation of the role and purpose of the Building Cost Information Service (BCIS on line 1998) in the provision and management of financial information.
- It provides a more readable and interactive text, using the supporting data contained in the website to demonstrate and support the concepts described and developed therein.
- It allows the reader to gain hands-on experience of design cost management and be made aware of its role and context within the constantly evolving design and construction process.

Acknowledgements

The authors of this book are particularly grateful to colleagues in many parts of the world, both in industry and academia, who have helped us to develop our ideas and understanding of construction economics and its role within the construction industry. We are also delighted to have had the opportunity to teach, for many years, students from various built environment disciplines, in many different countries, the subject of construction economics. Without their enthusiasm for the subject and their willingness to enter into discussions with us, we would not have been able to organise our thoughts and ideas and be able to present them, hopefully, as a reasonably coherent and comprehensive account of the subject of this text.

We are particularly indebted to Joe Martin, Executive Director of the Building Cost Information Service at the Royal Institution of Chartered Surveyors, without whose support and the provision of information from the service for use in the book and the accompanying website, neither could have been written. We are also very grateful to him for kindly preparing the constructive and supportive foreword to the book.

We are indebted to Tweeds for providing us with the comprehensive real-life project on which the case study, described in the book and demonstrated on the website, is based.

We are especially grateful for the contribution made to the development of the website by Ms Claire Ashton, of the School of the Built Environment at Liverpool John Moores University.

We also acknowledge the various sources of information which have helped us formulate our ideas which we hope we have fully referenced in the book. If any sources have been overlooked we apologise. Although we have endeavoured to eradicate as many as possible, any mistakes or errors in the book are entirely our responsibility.

Finally we would like to thank our families and friends for their support and encouragement when they hoped we might be doing something else with our time.

About the authors

David Jaggar MPhil PhD FRICS MCIOB MACostE is Emeritus Professor of Construction Economics and former Deputy Director of the School of the Built Environment at Liverpool John Moores University. He has many years of experience in industry and higher education as practitioner, consultant, teacher and researcher in the field of construction economics. He has published widely and given conference papers in many parts of the world; his consultancy has included work for the World Bank, North West Water and the Royal Institution of Chartered Surveyors. He is joint co-ordinator of CIB W92 Procurement Systems which is concerned with international building procurement issues.

Andrew Ross BSc MSc MCIOB MRICS is the head of the construction section at Liverpool John Moores University. He is a chartered surveyor and chartered builder and has many years of experience teaching construction economics to undergraduate and postgraduate students. He has published nationally and internationally in this field and in the area of information management. His research interests are in the areas of information and knowledge management in construction organisations and he is a consultant to a number of contracting organisations.

Jim Smith, MUP (Melb), PhD, FRICSIS is a Fellow of the Royal Institution of Chartered Surveyors. He has worked in private practice and local government in London, Brighton, Sheffield, Nottingham and in Melbourne, Australia. He taught at the National University of Singapore, the City Polytechnic of Hong Kong and the School of Architecture of Building at Deakin University, Australia. Jim is presently at the University of Melbourne and recently completed a PhD on strategic client briefing. He has numerous refereed international journal papers and refereed international conference papers to his credit. Jim's previous books, *Building Cost Planning for the Design Team* (1998) and *Building Cost Planning in Action* (2000), have been well received in Australia and overseas.

Peter Love BSc MSc PhD is an Associate Professor in the School of Management Information Systems at Edith Cowan University, Australia, and Asia Pacific Editor for *Logistics Information Management: An International Journal* and *Business Process Management Journal*. He has a wide range of industry experience which he gained in the UK and Australia working as consultant project manager and commercial manager for a multi-national contracting organisation. His research interests include supply chain management, quality management, IS project management and strategic information systems evaluation. He has co-authored/edited four books and has published over 200 internationally refereed research papers.

1 The context: definitions, historical influences and the basic approach

16 MAR

Introduction

This chapter aims to:

- Define some of the terms used in the processes of design cost management
- Introduce the reader to the historical imperatives for managing the costs of design
- Discuss the quantity surveying profession's role in the development of procurement documentation used for cost prediction and control
- Establish the basic processes for design cost management.

Definitions

When discussing the subject of design cost management, we often find that there are a number of terms and concepts that are used, either to mean the same as some other term or terms, and sometimes used out of context or even incorrectly. In fact the title of this book *Building Design Cost Management* has been deliberately chosen to fully reflect the various management processes that are necessary to go through in order to help the client achieve value for money. As a result of this decision the authors felt it important to identify and explain the various terms that are often used in association with the term 'cost planning' as an important requisite before undertaking a major study of building design cost management.

The term 'cost planning' is often used to reflect a process which is rather more embracing than its strict definition. So, starting from first principles we will define what the process is about; it is a total process concerned with the financial management of building projects by those responsible for the creation of the design and therefore it is about the management of the *price* the builder requires for the completion of the building project. We are not planning or controlling the cost to the builder, and for this reason, strictly speaking the term 'cost' is a misnomer. Nonetheless, for historical and embedded usage reasons we shall continue to use the expression, even though it is not entirely accurate.

Thus, it is about *planning* the likely financial implications of the project early in the design process and *controlling* the price during the development of the design, so that, the builder's price matches the initial financial plan. The term

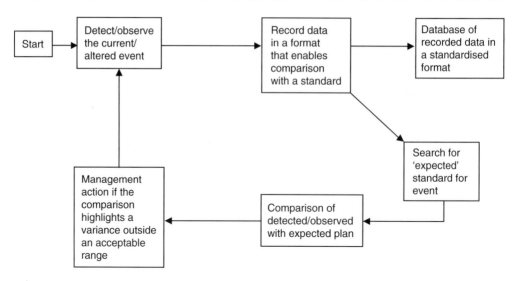

Figure 1.1 Management control system.

'cost planning', in reality, does not include the notion of control, which is an essential component of any financial management system (Fig. 1.1).

Another commonly used term to describe the above process, which has been used in the introduction to this book is 'design cost control'. Specifically this term reflects the process of ensuring compliance with the cost plan which is a statement of how much money should be spent on the various significant parts of a building.

Once the cost plan has been produced the next stage is that of cost checking to ensure that as more detail about the project is generated, the financial statements identified in the cost plan are complied with, and if not, the necessary remedial action put in place, to ensure that there is no deviation from that which has been agreed in the cost plan. Figure 1.2 shows the total system of design cost management as established since its inception in the 1950s.

Therefore, in truth, the term that should be used to fully describe the process of design cost control or cost planning as previously explained is 'design cost management', the meaning of each individual word being:

- design: the process of producing the required model of a building
- cost: the amount of money that the client expects to hand over to a builder, usually during and on completion of the building (the total price that the builder wants for completing his work)
- management: the responsibility for ensuring that the functions of planning, control and feedback are successfully brought about in terms of design, cost, time and quality (project objectives = project management).

Hence, as used in the title, the term 'design cost management' has been adopted throughout this book, to reflect the totality of the process although in practice we recognise that the terms 'cost planning' and 'design cost control' are often used to

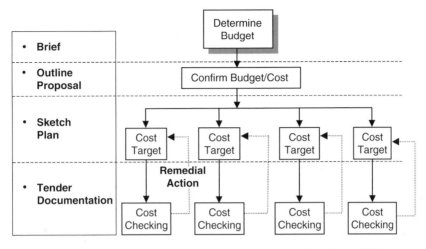

Figure 1.2   Design cost management concept map. Source: Smith and Love (2000).

describe this process. The various chapters in this book will explain, describe and demonstrate the application of the process of design cost management.

Fact file

There are various methods of procuring new construction work, all of which have their relative merits and demerits, which are addressed in Chapters 3 and 4, as it is important that the reader is made aware of how such changing trends impact upon design cost management.

   These changing trends are not considered in depth in these chapters. For a more detailed and comprehensive treatise of all the various procurement strategies highlighted, the reader is recommended to consult the literature (Aqua Group 1990, Masterman 1992, Morton and Jaggar 1995, Rowlinson and McDermott 1999).

A further term that is often referred to is 'cost modelling' which some experts in the field have identified as a separate subject. Much discussion and intellectual thinking has been attributed to defining, developing and promoting its use. Reference to cost modelling will be made later in the book but, at this stage, it is important to put the subject into context. A model is anything which represents reality, from a drawing, photograph or a plastic model of a Triumph Stag sports car, through to a mathematical expression or algorithm such as Einstein's famous equation $E = mc^2$. The reasons for using models are diverse and range from simply acting as a reminder of something, or someone with a photograph or facsimile model, through to a mechanism to explain a phenomenon, predict behaviour, or discover why things work in the ways they do! However, there is an interesting paradox in the use and application of modelling techniques which is that, generally the more abstract the model the more informative it will be in explaining behaviour, but the less easy it is to comprehend and therefore will tend to need specialist understanding in order to make sense. Hence mathematical models tend to be the most useful form of modelling, as they are symbolic representations of real life events and possess the essential requisite of being capable of manipulation in order to try out different scenarios. Going back to our Triumph Stag, our plastic model can occupy pride of place in our living room and serve as a permanent reminder of that particular car.

Alternatively, we may be able to build a complex computer-driven mathematical model reflecting the energy/performance characteristics of the engine. Such a model would hardly take pride of place in the living room, but would be vital to mechanical engineers wishing to learn about the car's power unit.

Turning to design cost management, the focus of this book, the reader can now see that the process is all about cost modelling, since at various levels of sophistication, we are trying to represent the likely financial implications of a real phenomenon, a building. In trying to model costs there is also a further paradox in that, when such models should be most useful, at the beginning of the design process, the less reliable such models are, as at that stage little mathematical description can be assigned to the project. Whereas when the design has been completed, a comprehensive mathematical description can be assigned, at a stage when, of course, it is of little use in seeking an optimal solution.

This book explains and demonstrates the various major cost modelling techniques that have been developed in order to bring about effective design cost management of building projects so the reader can be confident with regard to how they work and how they should be applied. It is also worth stressing that techniques alone, no matter how sophisticated, will only be effective if those using them possess the necessary experience and understanding of the implications of design and construction in terms of resources and costs. On another cautionary note, the forecast or prediction of future cost also reflects the quality and accuracy of the source data and its interpretation to the project in hand.

1.3   Historical development

Design cost management was first developed and deployed in the UK after the Second World War, when the ability of the quantity surveyor to predict the financial implications of yet-to-be-built building projects was becoming less reliable. Client bodies responsible for large building programmes (particularly central and local government) were becoming less satisfied with the role of the quantity surveyor who, up until this time, had been generally successful in estimating future building costs, at the early stage in the design process. Building materials shortages and a massive demand for buildings due to post-war rebuilding and the baby boom demanded improved methods for cost planning and control to ensure value for the public investment being made.

Why did this state of affairs occur? Probably this arose because of the expansion in the choice of materials and components, coupled with a greater variety of construction solution techniques, which meant that quantity surveyors' early attempts to predict costs, without any system of planning and control, were often unreliable. The problem was also exacerbated because of increasing inflationary trends during this time which again emphasised the need for a more effective system of planning and control.

Fact file

It was recognised by quantity surveyors in the formative years of cost planning that bills of quantities used in most building contracts at that time provided a rich source of cost information. They contained in a consistent and easily understood form the contractor's financial implications of a tendered building project. This was in terms of the quantity, specification and cost.

As a result of the pressure to improve the situation, the quantity surveying fraternity began to seek out possible solutions. They turned their attention to the rich source of data available in their own contract documents, bills of quantities. Bills of quantities, since they were introduced in the middle of the nineteenth century to help in the selection of contractors by competitive lump sum tendering, became essential contract documents for the following reasons:

■ They saved duplication of effort, as all competing contractors did not have first to extract their own measurements of how much construction work was required in the project.
■ All contractors tendered on the same information and therefore on an equivalent basis where each competitive tender bid could be more objectively compared.
■ They allowed for a more accurate, equitable and easily arrived at financial settlement of stage payments (generally monthly) as work progressed on site.
■ They facilitated the settlement of the final account, as inevitable changes occurring during the on-site stage could be financially assessed through the unit rates contained in the contract bill of quantities.
■ They allowed the contractor to use the information contained in the bill of quantities to assist in the on-site checking and management of the construction process.

As a result of these benefits, coupled with the massive increase in the need for construction of both buildings and infrastructure, to support the emerging Industrial Revolution in the mid-1800s, lump sum tendering, as a means of selecting a contractor to carry out the necessary work, became the accepted method of procurement. This procurement method presumed that the design and construction processes were sequential operations, based on full documentation and executed by different organisations, i.e. design by an Architect led team and construction by a Main Contractor led team. That is, specialisations were being established in the design and construction of buildings.

As a result of this success, the distinct profession of quantity surveying was established. It was underpinned by a Royal Charter and gained the status of a learned society, much of its contribution being based on bills of quantities, which became an integral feature of its activities. The bills of quantities achieved their success for the reasons already outlined, as they are based upon a simple but effective mechanism of representing 'descriptions of finished work together with their quantities'. This approach was so successful because of its effectiveness in communicating to another party; the tendering contractor and any other parties involved such as sub-contractors, what work was required and how much of it there was to be done. It also provided essential information in defining the resources needed, together with their financial implications: necessary requisites in the estimating and tendering process.

It was soon recognised by the quantity surveyors producing the bills and the builders using them, that the basic philosophy of units of finished work needed to be set in a more consistent framework so that better understanding, portability in their use and objectivity would be achieved. As a result the best practices of the day began to develop and apply sets of conventions to improve their production and use and thus their reliability. In 1922, after repeated attempts and many false dawns, the first English Standard Method of Measurement was produced (Standing Committee 1922). The Standard Method of Measurement is now in its seventh edition (CPI 1998), having being changed and updated approximately every 15 years or so since its inception. Although these various changes have taken place, perusal of the current method and the first method will show very little change in measurement principles, and in particular, the guiding philosophy, 'descriptions of finished work together with their quantities'.

Fact file

Bills of quantities had been developed in the middle of the last century as a basis to aiding the process of contractor selection using lump sum competitive tendering based on a completed design solution being supplied to a number of interested contractors. So successful did these documents become that the profession of quantity surveying developed based on being the custodians and producers of these documents. Formal rules of measurement were developed first in 1922 which were established and agreed between the quantity surveyors and the recipients of the bills as a basis for pricing the contractors and, more precisely, the estimators.

For a more detailed development of the history of surveying and quantity surveying the reader is recommended to study *Chartered Surveyors. The Growth of a Profession* (Thompson 1968).

In fact, as will be discussed later in this book, there has been much criticism of bills of quantities in that they do not model accurately the true costs of construction, since they only reflect approximately how costs are incurred and are, therefore, at best unhelpful to those concerned with the management of construction costs. Others contend that at worst they lead to erroneous interpretations of those costs. The simple fact remains that, despite these criticisms and the development of various alternatives, the bill of quantities still remains the prime source of cost information and still plays a significant role in lump sum procurement strategies.

So, having briefly explored the development of quantity surveying and bills of quantities, we can now explore the role these documents played, and still play, in the establishment of design cost management. As we suggested earlier, quantity surveyors were unable to reliably predict the financial implications of future building projects so new approaches were necessary. As with most good ideas there was no blinding flash and eureka but a series of progressive and linking developments, which continue today and will no doubt evolve in the future.

The first step was in answer to a problem that was occurring in the 1950s in that central government was responsible for funding local authority projects throughout the country. After the war there was rapid expansion of housing and education. As a result there were problems in terms of securing value for money, as costs of similar projects varied considerably between different local authorities for no obvious reason.

It was realised that by analysing bills of quantities for similar historic projects it would be possible to gain a better understanding of the underlying costs of the schemes. The basis for this approach was the realisation that what was needed was a device to express costs in a meaningful way that could be used for comparative purposes.

Fact file

Bills of quantities became so successful for the following reasons:

■ saves duplication of effort: all tenderers tendering on same information
■ greater objectivity in tendering process: all tendering on same information
■ basis for valuing work as work proceeds, especially monthly valuations
■ basis to aid valuation of work during final account as inevitably changes will occur.

As a result of this need elemental cost analysis was developed so that one project could be compared with another in a consistent way. So what is elemental cost analysis? Why is it so useful? It was and still remains the best means to achieve a re-allocation of a priced bill of quantities from a work section, or, earlier, a trade format, into an elemental format. The major difference in presentation is that work section formats represent a bringing together of similar kinds of work, such as all the brickwork or concrete work. This helps facilitate the estimating process by the contractor and in turn, the subcontractor, bearing in mind that the time available for the tendering process to be carried out was, and is, often very limited.

Fact file

Bills of quantities used by contractors also aid financial management, including planning and control, of the task of construction. However because essentially the unit of measurement is that of finished work this use is somewhat limited and not entirely accurate as contractors utilise resources to achieve the finished product. This issue has been much debated and various attempts have been made to try to enhance the relationship between the product of construction and the process of construction but there is still a considerable distance between the resolution of this issue although SMM7 (CPI 1998) rules did move someway towards this by the recognition and incorporation of method related charges recognising non-quantity related information in the Preliminaries to bills of quantities. Reference will be made to this later.

However, such an arrangement for studying building costs by the construction economist was not entirely helpful so the trade or work section bills were re-allocated into an elemental format which allowed easier and, more importantly, more accurate comparisons to be made. So what is so useful about elemental cost analysis? Consideration of the definition of an element will help explain. It can be defined as:

> The major parts of a building that always play the same function irrespective of their location or specification ... (Ferry *et al.* 1999).

Or, which is essentially the same:

> ... a major component common to most buildings which usually fulfils the
> same function, or functions, irrespective of its construction or specification
> (BCIS 1969).

Elemental cost analysis is the perfect vehicle for making comparisons between one
building and another and being able to draw useful and reliable inferences from
such comparisons. It also provides a common functionally based means of communication between members of the design team (and the client). An example will
serve to illustrate the point. Consider the concept of a roof, with its attendant
structure, coverings and drainage as described by the BCIS in its List of Elements
(2C) as shown in the Standard Form of Cost Analysis (SFCA) shown in Appendix A2. We can see the roof is an element as it meets the definition stated above.

Consider two typical cost analyses for building project A and building project
B for the element of roof. Building project A has a pitched roof in best quality
slates with an area of $5000\,m^2$ and a total cost, as extracted from the bill of quantities, of £150 000, whereas building project B has a flat roof with one layer of felt
and an area of $500\,m^2$ and a total cost, as extracted from the bill of quantities, of
£5000. There is a considerable difference between the two costs that needs an
explanation.

However, we can quickly see why this is so when we start to investigate at even
the most basic of levels. We notice that there is a considerable difference in the
quantity of work involved and also in the specification. So we can account very
rapidly for the differences in cost. We can also begin to identify other obvious differences. Perusal of the drawings will tell us how complex is the layout and
arrangement of each roof. One roof may contain many discrete areas and/or complex arrangements. Also, perhaps the contractors were not pricing the bills during the same time period. Costs and prices will vary at different time periods. So,
clients today would expect to pay somewhat more for the same building than they
would have 20 years ago, or even 20 months ago. Further perusal of the two projects will also reveal other relevant information like their location. For example,
is the site in the middle of central London or in a rural location in the north of
England? Again, the location would influence the project cost and the costs of the
various elements

One might argue that exactly the same information can be gleaned from the
original work section tendering bill. This is quite true. The only difference is that
it is the same information, but rearranged to make it more interpretable. However, the secret is that the rearrangement must take place to enable like to be be
compared with like. To illustrate the point, when comparing the work section of
brickwork between projects it will tell us how much there is and its specification,
but not what it belongs to, for example external or internal walls, foundations, or
part of the roof. Each of these different locations can influence costs which may
be reflected in the contractor's unit rates but of course the work section presentation will not tell us to which of the various elements and in what quantities the
brickwork belongs to. We can now see why elemental cost analysis became so useful
as the same means of comparison.

So, armed with this technique, the quantity surveyors were able to explain
why a school in Cambridge, for instance, cost more or less than a similar school

in Liverpool and, therefore, were better able to enter into constructive dialogue with the designers, engineers and local authority clients responsible for their provision.

Fact file

During the 1950s a connection was made that it might be possible to reprocess the information contained in the priced bills of quantities into functional breakdowns termed elemental cost analyses which gave a better understanding of what made up the costs of buildings. Initially the technique was used to find out or analyse why a given building cost what it did rather than as a technique to predict the costs of future buildings. The use as an analytical technique arose initially because of central government wishing to understand better why buildings, especially those concerned with the rapidly expanding housing and schools programme, seemed to cost more or less in different parts of the country to try to ensure that proper value for money was being obtained on behalf of society being served by the various local governments.

It was then quickly recognised that this technique could be used to predict the future costs of buildings and the technique of design cost management based on elemental cost analysis was established.

It was a relatively short step to the realisation that the technique of using elemental cost analysis to explain why buildings and their elements cost what they did, could be used as a predictive tool. That is, it could be used to ascertain the likely costs of a future building with more certainty, to overcome the shortcomings of the existing prediction techniques as previously highlighted. These limitations would be overcome since the developing technique would allow for the necessary remedial action to be taken during the evolving design process, such that, when the lowest bid submitted by the contractor was received by the design team, they would be reasonably confident that it would equate closely with how much the client should expect to pay in their budget, as advised by the design team, at the commencement of the scheme.

It is also important to stress that, although we are talking about developments that occurred 40 years or so ago, the basic principles, which are developed later in this book, remain relatively unchanged. The basic techniques have been subject to refinement and the application of developments in information technology, rather than there being any major changes or innovations.

1.4   The basic approach

The technique of design cost management was established on the foundation of elemental cost analysis largely based on the analysis of tendered bills of quantities. So how did the process work? It worked on the philosophy of predicting or planning how much the building was likely to cost at a very early stage in the design process by simple, yet effective methods of estimating. Thus, the budget for the building project was established. Incidentally, it was at this stage where the previous prediction techniques stopped and thus no remedial action was possible before the submission of the tender. Where such disparities did occur, as was often the case, much abortive work would be required in order to ensure that the tender bid matched more closely the original budget. Also, often of greater concern, was the fact that such abortive work often led to inappropriate cost savings being made, as a more fundamental approach would have meant literally 'going back to the drawing board'. Such 'savings' often embraced the less disruptive substitution of cheaper finishes or lower specifications of materials and components, which invariably would lead to more expensive maintenance and running costs being incurred as well as the likelihood of less satisfactory user satisfaction being achieved.

Fact file

Reasons for development of design cost management: it was quickly recognised that for the system to become reliable and helpful a degree of consistency was required together with guidelines as to how best to carry out the process. As a result a 'think tank' was established by the Royal Institution of Chartered Surveyors in the 1950s and the idea of design cost management based on the use of elemental cost analyses derived from priced bills of quantities was given formal backing. Support was also given to the establishment of a nationally agreed set of conventions for the establishment of cost analyses together with the linking of the process to the RIBA plan of work (RIBA 1998).

As a result a standard form of cost analysis was established which laid down the following:

- definitions of terminology
- the elements and the definition of each one
- how to carry out the preparation of the analyses from priced bills of quantities.

It is important to distinguish between *designing to a cost* and *costing a design*. The former means working within a budget established by the client at the outset and tailoring functional, technical and quality requirements to suit the defined budget parameter. In costing a design the client needs and requirements are first established and the role of the design team is to establish a realistic and affordable budget to provide these stated requirements. These two processes are summarised in Fig. 1.3.

The notion of being able to explore various solutions is an important attribute of design cost management. It means that, although a budget has been set at a very early stage in the design process, it does not limit the various solutions that might be available for exploration, as the production of the cost plan and the execution of the cost checks will ensure flexibility, as well as confidence, in the development of the final design and construction solution, because, as will be demonstrated, design cost management is a dynamic process, something which was not inherent in the earlier cost prediction techniques.

Figure 1.4 illustrates the fact that it is at the earliest stages in the design of a building where the greatest opportunity to investigate a variety of solutions is possible. It also shows that opportunities for savings later on become much less available.

Once the budget has been established, the next stage is producing a cost plan. At its simplest level this process requires the following steps to be taken:

(1) Find a suitable cost analysis of a similar building project: the closer the match the more successful the design cost management process is likely to be. The match should be in terms of building type or function. If we are building

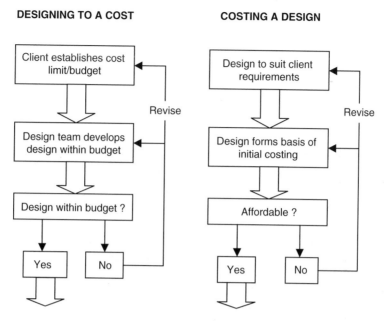

Figure 1.3   Basic approaches to design cost management.

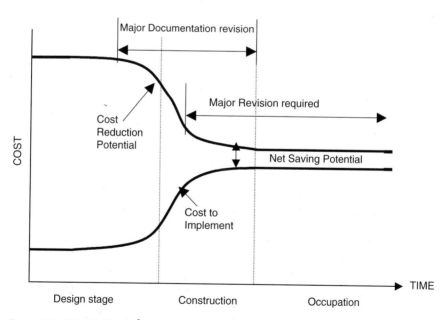

Figure 1.4   Potential for savings, adapted from Flanagan and Noman (1963).

a school then we need to find an analysis of a school, preferably of the same function, that is, both secondary schools. The reason for seeking such a match is reasonably obvious in that generally the specification of the materials and components, as well as the general design parameters, will be fairly similar.

Fact file

A more subtle argument that actually questions the whole philosophy of the tendering process is the use of what is termed 'social opportunity costing'. Here the tendering contractors actually build up their tender bids on what the market expects to bear. An interesting experiment was conducted some years ago (Fines 1974) where one set of tendering contractors was asked to prepare bids for a building shell described as a barn and another set of contractors was asked to prepare bids for the same shell but identified as a repertory theatre. Remarkably the bids for the theatre were in the region of 10 times that of the bids received for the barn (£3 000 000 for the theatre and £30 000 for the barn). This reinforces the need to match building type or functions and highlights the problems associated with market driven prices as will be discussed later in Chapters 3, 5 and 12.

(2) In identifying a suitable function or type the closer the match the more likely the success of the design cost management. Matches in terms of the following should be sought:

- size and general arrangement
- number of storeys
- complexity of plan shape
- nature of the ground both in topography and geology.

Clearly all these factors have a bearing on the resource and financial implications of a building project, in terms of the specification, the design solution and ultimately the construction solution. For example multi-storey construction necessitates staircases, perhaps lifts, possibly a frame, heavier foundations to accommodate the heavier loads, etc., whereas single-storey solutions to enclose the same floor area will need greater external walling. Also, more subtle considerations come into play such as for the same site plan a multi-storey solution will necessitate more external works to be carried out as there will be more site area to develop outside that occupied by the building footprint than an equivalent single-storey solution.

Fact file

It is perhaps worth making the point at this stage that the process of design cost management, and all the techniques used to bring it about, such as those being described here, is very much dependent upon the experience of the construction economist. To that end the context of the particular project under consideration must always be taken into account when giving cost advice. An example, that one of the authors experienced, which demonstrates the need to think, not just about the building project but its relationship to the general infrastructure within which it is to be located, occurred whilst on a visit to Riyadh in Saudi Arabia. Although there appeared to be an abundance of land upon which to build, and bearing in mind that generally the taller the building the more expensive it is likely to be, all the buildings were multi-storey solutions. Why? The answer on further investigation proved to be simple: basically there was no infrastructure in terms of roads, mains sewage, electricity, etc., except within a confined area in and around the city. Thus, the only realistic solution was to minimise the footprints of the buildings by building multi-storey solutions.

(3) Select an analysis which represents a building which has been completed as recently as possible, because although it is possible, through the use of indices, which measure changes in prices and costs over time, to adjust for these differences, it is likely that the greater the time span involved, the less reliable such indices will be, due to changes in technical solutions, fluctuating market conditions, client requirements and expectations.

(4) Select an analysis which is as close as possible in terms of geographic location. Such locations will tend to break down into macro and micro considerations. For example it is generally less expensive to build in the north of England than the south of England but of course the nature of the site in terms of ease of access, loading and noise restrictions, planning impositions, etc., will have a profound effect on the likely building costs. For example to build in central Mayfair in London will be more expensive than in an open country location. Again there are indices available to measure differences in the level of costs and prices in different locations. These aspects will be developed later in the book.

Fact file

Procurement strategies and context setting
Elemental design cost management developed due to the problems of early cost prediction within lump sum competitive tendering strategies and paradoxically it was this very process, based invariably on the use of priced bills of quantities, as described above which provided this rich database under the custodianship of the BCIS and also as developed within a number of practicing quantity surveying environments. Such tendering strategies were invariably the norm due to: the well established procedures for their management, inertia within the industry, the influence of the professions, education, public accountability and ease of implementation.

However, in later years, initially through client dissatisfaction with the approach there was considerable demand for improvement led by initiatives such as the British Property Federation (BPF) system (1983) and in the USA moves towards the use of management and construction management contracts. Such client demand has led us to the situation we have today where design and build and other alternative forms of procurement are becoming more predominant.

Thus, the technique of cost planning, as a part of design cost management, was and remains, to select an analysis (or analyses) as previously described above and adjust it (or them) for:

■ time
■ location
■ quantity
■ quality.

In this way we can produce a cost plan or cost model for each element of the proposed building. Once we have completed the cost plan we should confirm the original budget and, if no further refinements are necessary to the various elements in terms of quantity, arrangement and specification, it can be confirmed. This can take some time and several attempts as it is an iterative process which must be undertaken until the design team is satisfied that the budget and cost plan are fully reconciled.

This all sounds deceptively simple but in practice it may not be quite so easy. The main problem encountered, and even more during the formative years of design cost management, was the lack of a suitable cost analysis from which to build the cost plan. In fact a further paradox can be identified which is, that the more out of the ordinary the proposed building under consideration is, the more difficult it is to predict its likely cost implications, and the less likely it is that any suitable cost analyses will be available. It is also possible that more than one analysis might form the basis of the cost plan, as different analyses may contain appropriate and relevant information to help in building the cost plan. This approach is supported by Flanagan and Tate (1997) and illustrated in Fig. 1.5.

Again such approaches are fraught with difficulty in achieving consistency and thus objectivity when building the cost plan. How such difficulties can be taken account of will be addressed later.

The final stage of the design cost management process is perhaps the most objective part of the process, cost checking. This part must be carried out in order

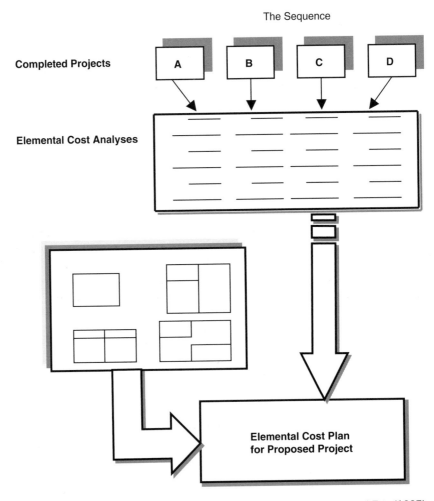

Figure 1.5   Essential elements of elemental cost analysis. Source: Flanagan and Tate (1997).

to confirm that the budget and cost plan are accurate. This is the cost control part of design cost management. It is carried out by replicating how the contractor would price the completed bill of quantities except we use a simplified form of measurement and pricing. This technique is called approximate quantities which are established from the developing design solution, by the construction economist who then allocates unit rates. The establishment of the unit rates is an expedient in that minimal reduction in accuracy is sacrificed in exchange for a substantial reduction in the time that is needed to prepare and price such approximate quantities when compared with the detailed quantities which are required under the rules of the Standard Method of Measurement. In fact, some would argue that the approximate quantities approach to measurement should be used more commonly in tender documentation; especially for those procurement methods with greater contractor responsibility and involvement in the design stages.

The reason that these approximate quantities can be prepared in the first place, is that the design solution is expanding in terms of graphical and specification information, which of course facilitates their preparation. The main point of all this is to ascertain whether the cost target that has been set against each element in the cost plan is correct or not. If it is not then the necessary adjustments can be made to the elemental targets set in the cost plan, to ensure that the budget remains valid. The final arbitrator of whether or not the budget and cost plan confirmed by the cost checks is correct, will be the accepted contractor's tender bid which, all things being equal, should equate with the budget. Clearly, if there are disparities between these two figures, then the design cost management, carried out for the project has been less than satisfactory. It is important to note that a tender bid falling below the budget is as unsatisfactory as one rising above as it means that a less than optimal solution for the project will have been established, as more 'building' could have been provided.

1.5   Summary

The process of design cost management has now been outlined. It is fully described and illustrated later in the book and the process further demonstrated by reference to the website. However, before moving on, it is worth reminding ourselves of the main aims of the process.

The aims of design cost management can be summarised as:

- To set a realistic budget at the early stages of the design process.
- To seek value for money by facilitating various design solutions to be considered thus ensuring that the cost targets set for each element represent best value for money within the context of the total building project. For example, there is no point designing a roof covering with a life of 100 years if the project is designed to have a life of 30 years.
- To provide a means of taking remedial action through a system of control, such that when the tender bid is submitted, it will equate closely to the budget set for the project.

A summary of the dynamic and iterative process of design cost management was shown in the concept map of the process in Fig. 1.2. This structure should be

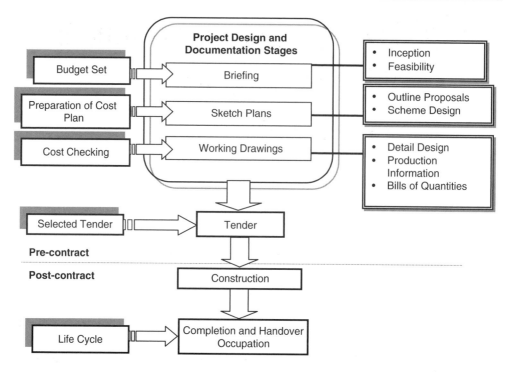

Figure 1.6   Stages of a project.

retained for all the work in this book as it will be used to illustrate the development of design cost management within the context of the maturing design solution as represented in the project documentation. Figure 1.6 shows the design cost management process within the total construction project life.

1.6   Reader reflections

- How might the construction economist's approach change if the design team were developing a design to a pre-determined cost?
- Why has the information contained in project procurement documentation got a value to the construction industry at large?
- What do you understand by the basic approach to design cost management?
- Why do you think that some construction economists consider that contractors produce 'socially acceptable' estimates rather than 'true' cost estimates?
- How may the Standard Method of Measurement affect the collection of cost information?

2 Design cost management: the cost planning infrastructure

Introduction

This chapter aims to:

- Discuss the development of the strategic cost information service provided by the BCIS
- Identify the various models in use that are used to reflect the design process
- Consider the influences of educators
- Introduce the reader to the principles of costs in use or life cycle costing.

In establishing the process of design cost management it was soon realised that two essential improvements were required if the technique was to achieve its full potential, namely:

- There was the need for a standardised approach to the preparation of the cost analyses in order to increase their reliability and portability.
- There was the need to establish a comprehensive library of cost analyses, preferably held by a national body, to be maintained by and to be used and accessed by the quantity surveying profession.

The first response was that the Royal Institution of Chartered Surveyors (RICS) set up the Building Cost Information Service (BCIS) which produced, on behalf of its members, a SFCA (BCIS 1969), together with guidelines as to its use, in order to prepare elemental cost analyses from priced bills of quantities. Details of the use and application of cost analyses is described later, but it is sufficient to say here that the SFCA did much to resolve issues of inconsistency and misunderstanding in terms of the preparation and use of these analyses as essential aids to the design cost management process. The development of the SFCA and the establishment of the BCIS on an information-sharing basis was a far-sighted initiative that we in the expanding *Information Age* are now able to reap the full benefits of such a visionary legacy.

The BCIS was set up, principally, to act as custodians of the various cost analyses, so that such a library would serve the needs of this developing and increasingly significant form of cost advice. Many quantity surveying organisations had already begun to produce their own libraries of cost analyses, as, generally, it was considered that information produced in-house would be more useful and reliable than that from an external source. Nevertheless, even a large concern such as the BCIS has difficulty in establishing a comprehensive library of such analyses. It requires a considerable commitment and investment in time, personnel and office resources. In fact, it is so difficult that it is virtually impossible, whereas collecting centrally, on behalf of the profession, and sharing the burden, clearly provides the potential for a much more comprehensive database. The BCIS also began to produce and develop various indices, as referred to earlier, as well as giving information about economic activity and trends, providing detailed studies of construction related activities and providing abstracts of various publications, relating to construction economics.

So, the BCIS at the present time offers the following range of services:

- elemental analyses of all types of buildings
- average building prices, element prices and functional prices
- tender price indices, output price indices and cost indices
- forecasts of cost and price trends
- location factors and regional price indices
- average preliminary percentages
- average contract percentages, profit on nominated subcontractors, etc.
- detailed briefing on the construction economy
- construction statistics, market reports, outputs and orders, etc.
- economic indicators: retail prices, interest rates, etc.
- daywork rates
- wage agreements.

The service, which continues to be funded through annual subscriptions, relies on the maintenance of its key feature, the library of cost analyses. Such analyses, prepared in accordance with the SFCA, are prepared and made available by the custodians of the priced bills of quantities, the quantity surveying profession. Further reference will be made to the BCIS as well as to its sister development, The Building Maintenance Information Service (BMIS) (BCIS on line 1998), which provides a similar service, but deals with building maintenance and running costs.

It can now be seen that there is an established infrastructure in providing design cost management, supported by a nationally accessible service concerned with providing cost data, source material and information to underpin this expertise. Despite this impressive progress, further enhancements still needed to be put in place. The first was to formalise its role within the various stages in the development of the design process, under the predominant means of tender selection in use, competitive lump sum tendering.

Thus, the premier management tool that was developed by the Royal Institute of British Architects, the RIBA plan of work (RIBA 2000), in fact, incorporated the various design activities to be carried out in addition to the cost control activities stages. Figure 2.1 shows the RIBA plan of work and the role of design cost control within it.

Figure 2.2 illustrates how the total cost control process fits in with the RIBA plan of work.

STAGE	AIM	MAIN PARTICIPANTS	USUAL TERMINOLOGY
A. Briefing: Inception	■ To prepare general outline of requirements and plan future action.	Client interests, architect.	**Briefing**
B. Briefing: Feasibility	■ To provide client with an appraisal and recommendation ... (to) ... determine the form ... (of) ... the project ... ensuring it is feasible functionally, technically and financially.	Client representatives, architects, engineers and quantity surveyor ...	
Outline Proposal can only begin after the architect's brief has been determined in sufficient detail.			
C. Outline Proposal	■ To determine general approach to layout, design and construction ... (and to) ... obtain authoritative approval of the client.	All client interests and consultants	**Sketch Plans**
D. Scheme Design	■ To complete the brief and decide on particular proposals, including planning arrangement, appearance,constructional method, outline specification and cost, and obtain all approvals.	All client interests, design team and approving authorities.	
Brief should not be modified after this point.			
E. Detail Design	■ To obtain final decision on every matter related to design, specification, construction and cost.	All design team	**Working Drawings**
Further change in location, size, shape or cost after this time will result in abortive work.			
F. Production Information	■ To prepare production information (tender documentation or working drawings) and make final detailed decisions to carry out work.	All design team	**Working Drawings**
The process continues into the tender, construction and hand over stages.			

Figure 2.1   RIBA plan of work. Source: adapted from RIBA (2000).

It is also worth pointing out that there are other models that have been developed which identify the different stages in the design process and the tasks to be carried out during those stages. Five such models that have been developed in Australia, the USA as well as in the UK are shown in the 'Fact file'.

Fact file

Design stages in three countries

Stage	Organisation				
	RAIA[1] (Australia)	NPWC[2] (Australia)	CIDA[3] (Australia)	RIBA[4] (UK)	AIA[5] (USA)
Briefing	Pre-design (briefing) site analysis	A. Briefing: Inception feasibility	Phase 1 Project initiation	A. Inception B. Feasibility	1. Pre-design 2. Site
Sketch plans	Schematic design/ development	B. Outline proposal	Phase 2 Project planning and design	C. Outline proposal	3. Schematic design
	Design development/ development	C. Sketch design		D. Scheme design	4. Design development
Working drawings	Design development/ development Contract documentation	D. Tender documentation	Phase 3 Procurement (and construction)	E. Detail design F. Production information	5. Construction documents

Sources: [1]Royal Australian Institute of Architects (1993); [2]National Public Works Conference (1993); [3]Australian Institute of Project Managers (1995)/Construction Industry Development Agency; [4]Royal Institution of British Architects (2000); [5]American Institute of Architects (1987).

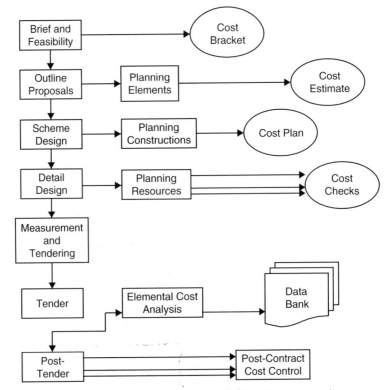

Figure 2.2 The RIBA plan of work adapted to show design cost management.

2.2   Technological development

Despite these advances there were still problems concerned with the production of the cost analyses themselves, in that the task of converting the work section bill into an elemental breakdown was time consuming and error prone. Initially many quantity surveyors began to produce elemental bills of quantities in order to avoid this difficulty. Unfortunately they proved to be unpopular with the tendering contractors because they made their task of estimating and tendering, in a short period of time, more difficult, due to the expansion and fragmentation of similar work throughout the bill. This conflict of interest between the need for elemental breakdowns by the design team and the work section breakdowns by the tenderers, began to be resolved in the 1960s with the introduction of mainframe computing into many local and central governments together with some large design practices to help to manage their accountancy-related functions. At this time there had also been improvements in the reliability and portability of bills of quantities by the introduction of Libraries of Standard Descriptions for bills of quantities, the most successful being that produced by Fletcher and Moore, a large, private quantity surveying practice (Fletcher and Moore 1979). It was quickly realised by the quantity surveyors, especially those working in local government environments, that the mainframe computer could be harnessed to produce, with speed and accuracy, different bill formats for different needs including elemental and work section breakdowns, thereby overcoming the difficulties highlighted earlier. In fact a very successful local government consortium, LAMSAC, was established which, amongst other things, developed a computerised library of standard descriptions (Local Authority Management Services and Computer Committee 1970), derived from the Fletcher and Moore library, which, together with the relevant software, facilitated the production of bills of quantities in different formats.

Figure 2.3 shows the range of bill formats that could be produced by the computerised LAMSAC system developed in the 1960s and 1970s. It demonstrates the range and power of the computer to prepare differently formatted bills of quantities if the measurement data have been coded appropriately in the working drawings/measurement stages.

Unfortunately, the system was based on expensive mainframe systems often housed in their own spacious facilities and 'owned' by the Council Accounts and Rating Departments. They were difficult and cumbersome to use as they often relied on 'punch cards' with overnight and weekend batching of jobs. So, gaining access and communicating with the machine was tedious, inflexible and frustrating. However, rapid developments in hardware, first through mini and then microcomputers vastly changed the environment for bill production and other contract documentation. The universal and personal computing power provided through laptop and desktop personal computers has expanded opportunities and possibilities in computing exponentially. These technologies are continuing to develop, for example those based on mobile telephone technology, which will provide even more flexible working environments at even lower costs.

LIBRARY

class no. 692·5 JAG

25181

BILL CODE	NO.	BILL TYPE DESCRIPTION
BLANK	1	• WORK SECTION
E	2	• ELEMENT/WORK SECTION
F	3	• FEATURE/WORK SECTION
B	4	• BLOCK/WORK SECTION
S	5	• BLOCK/SUB-BLOCK/WORK SECTION
EF	6	• ELEMENT/FEATURE/WORK SECTION
BE	7	• BLOCK/ELEMENT/WORK SECTION
SE	8	• BLOCK/SUB-BLOCK/ELEMENT/WORK SECTION
BF	9	• BLOCK/FEATURE/WORK SECTION
SF	10	• BLOCK/SUB-BLOCK/FEATURE/WORK SECTION
BEF	11	• BLOCK/ELEMENT/FEATURE/WORK SECTION
SEF	12	• BLOCK/SUB-BLOCK/ELEMENT/FEATURE/WORK SECTION
BWE	13	• BLOCK/WORK SECTION/ELEMENT
SWE	14	• BLOCK/SUB-BLOCK/WORK SECTION/ELEMENT
BWF	15	• BLOCK/WORK SECTION/FEATURE
SWF	16	• BLOCK/SUB-BLOCK/WORK SECTION/FEATURE
BWEF	17	• BLOCK/WORK SECTION/ELEMENT/FEATURE
SWEF	18	• BLOCK/SUB-BLOCK/WORK SECTION/ELEMENT/FEATURE
WE	19	• WORK SECTION/ELEMENT
WF	20	• WORK SECTION/FEATURE
WEF	21	• WORK SECTION/ELEMENT/FEATURE

Figure 2.3 Typical bill formats produced by the LAMSAC computer system.

The great benefit of the bill is that it provides a third party with an instant and easily recognisable view, as to precisely what is required by the client in order to construct the building, quantities and descriptions of finished work, couched within an agreed framework. Unfortunately, this significant benefit does little to aid the process of construction on site, as all the various resources required (materials, components, labour, plant and supervision) are locked up within these units of finished work. Additionally they tend to ignore where the work is actually to be carried out, except on a very broad level. This means that the contractor has the difficult task of manipulating the information contained in the bills into a model, to reflect how the project is to be put together on site, by creating a programme. Clearly this means much disaggregation of the information in the bill in order that it can be reassembled into essentially a time-driven model. This problem, difficult as it is, also has the added disadvantage that the contractor has already priced this information at the tender stage. Such pricing clearly contains a series of average prices, as different locations of the same work tend to be aggregated. To give a very simple example, the top identical floor of a multi-storey building will clearly be more expensive than an identical floor near the bottom, but such differences are not directly reflected in the bill of quantities.

This difficulty has been recognised for many years and a variety of proposals have been put forward to resolve it including the Operational Bill (Forbes and Scoyles 1963), the BPF system (British Property Federation 1983), Construction Planning Units (Property Services Agency 1969), and the Building Industry Code (CLASP 1969). Improvements have also been made in the latest edition of the Standard Method (CPI 1998) some of which were developed from the Civil Engineering Standard method (Institution of Civil Engineers 1985) which recognise and allow for the separate financial consideration of resources with low deployment flexibility such as a tower crane, the costs of which are ongoing, whether or not it is used on site, and non-quantity related costs such as bringing expensive plant to site and its subsequent removal, the costs of which have nothing to do with how much work that plant does between these two stages.

Equally dramatic advances in software, much of it generic and universal, have supported these developments and innovations. This means that all users have extensive computer power at their fingertips and it is simple, easy, reliable and entirely global, through the expansion and development of the World Wide Web. The use of these developments will be discussed later. At this stage, it is sufficient to say that information technology has played a critically important role in helping to provide momentum for the establishment and development of design cost management. In fact, quantity surveyors have always been at the forefront of computer developments in the construction industry, starting way back in the 1960s with the primitive first forays into computer-generated bills of quantities described earlier. Quantity surveyors have been keen proponents of the new technology and computer-generated developments. This characteristic has ensured the survival, evolution and expansion of quantity surveying services when external observers had written them off in each generation.

Design cost management is not an end in itself. Its aim is to contribute to better buildings that give value for money to the client, users and the community. Design cost management, design cost planning and design cost control are an essential part of the design team's armoury of skills, knowledge and techniques.

So, the history and development of design cost management and the various influences that have led to its establishment, are essential in trying to gain for their clients, the best value for money.

2.3   Educational influences

Before we go on to look in detail at how the process works, there are still a few other important influences that need to be highlighted. Education, especially through the introduction of degree courses in the early 1970s, did much to introduce and encourage innovation when both educators and students began to explore more fundamentally the role of the quantity surveyor and how other techniques and ideas could be applied, such as information technology and the use of mathematical modelling techniques. All these developments began to have a significant impact on the process of design cost management and most graduates from such courses were equipped with a much broader understanding of what it was all about, together with an awareness of some of its limitations. Perhaps the best example of this was the development, as part of a funded research project undertaken by the Department of Surveying at Salford University and the Royal Institution of Chartered Surveyors, Quantity Surveyors Division, of an Expert System (ELSIE 1988) which had, as part of its application, a module designed to produce a cost plan based very much on the techniques described earlier and to be addressed in more detail later. As part of an Expert System the production of the cost plan was by professional expertise, captured within its inference engine, from quantity surveyors with recognised expertise in design cost management.

Figure 2.4 shows the four component parts making up the Expert System arrangement; budget, procurement, time and development appraisal.

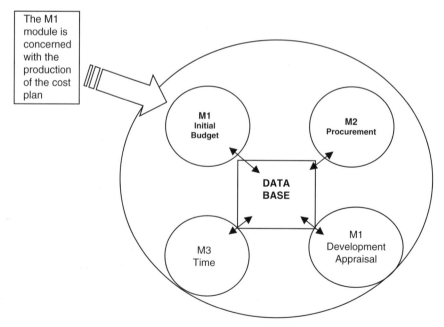

Figure 2.4   Overview of the ELSIE Expert System (adapted from ELSIE 1988). Source: Quantity Surveyors Division, The Royal Institution of Chartered Surveyors.

Alongside these developments, the BCIS in 1984 (BCIS 1984) supplemented its hard copy service (where the various subscribers receive their updates by mail) by introducing access to its database, including all the analyses, via a telephone modem link.

This proved to be a very exciting development and made the use of the system much more accessible, dynamic and instant (on-line). In addition to being able to access the database, the BCIS also introduced some modelling techniques which provided the following support for the quantity surveyor:

- Statistical manipulation of overall building costs such as determining means, modes, standard deviations, etc., coupled with the ability through its indices to reflect changes in cost and price due to time and/or location.
- The introduction of an Approximate Estimating Package, which allowed the quantity surveyor (having identified suitable cost analyses from which to build a cost plan) to prepare on their own computer, a cost plan for the proposed building project. A great advantage of this development was clearly the ease and interactive nature of the cost planning process with the database. It allowed the user to carry out with speed and accuracy, as much iteration as necessary in order to seek the most effective design–cost solution.

The latest in a long line of developments at the BCIS occurred in 1998 (BCIS on line 1998), with the establishment of a website address, where subscribers access and obtain the information needed for their design cost management requirements. The aim of this book and the accompanying website is to demonstrate its use in a modern web-based approach to the design cost management process. With the development of the website it was decided by the BCIS to abandon, for the immediate future, the Approximate Estimating Package, as it was felt that in-house cost modelling applications, customised by the subscribers, based on spread sheet applications which interface with the service's database, would be a more useful direction to pursue, rather than trying to produce a generic application, such as the Approximate Estimating Package.

2.4   Life cycle costing

Before going on to discuss how today's discipline of design cost management is carried out, and the role it plays within the building industry it is worth mentioning some other relevant advances which are concerned with bringing better value for money to our clients and society generally.

The use of the building during its life has enormous financial implications, which are generated and committed in its design, and therefore, this should be considered by the design team at the earliest possible moment in the design process. Typical examples are ensuring that the specification of the materials and components match the anticipated life of the building. There is little point in specifying a long lasting floor finish if the building is designed to have a short life and of course the opposite is true. Of particular significance are energy and running costs, which need considerable thought and consideration at the design stage.

A whole discipline of life cycle costing has been established over many years, underpinned by a similar service to that of the BCIS called the BMIS (BCIS on line 1998).

Fact file
A definition of life cycle costing is 'the present value of an asset (a building in our case) over its operating life, including initial capital cost, occupation costs, operating costs, and the cost or benefit of the eventual disposal of the asset at the end of its life. Life cycle cost techniques take into account, during the design and management of construction projects, the total costs that the project will impose upon the client during the whole of its life' (Quantity Surveyors Division of the Royal Institution of Chartered Surveyors 1986).

Figure 2.5 illustrates dramatically why life cycle cost studies are so important in the establishment of the budget and cost plan for a future building. There are numerous excellent texts which describe and illustrate how life cycle costing can be carried out (Dell'sola 1981, Flanagan and Norman 1983). However, one of the major difficulties is the regular collection of the data due to the many years of life of a building. The reliability of some of the data can, therefore, be questioned.

A more recent development has been the introduction and expansion of the discipline of facilities management which recognises the need to integrate the requirements of the organisation using the facility or building with its operation. Whilst facilities management is not the primary focus of this book, the role of the BMIS (BCIS on line 1998) is to support decision-making in facilities management, and for our purposes it is to recognise that the running and operational costs (the hidden life cycle costs in Fig. 2.5) are largely dependent on the design decisions made during the formative stages on the life of a project. Therefore, the design team (as well as facilities managers) need to be fully aware of the content and life cycle data that BMIS offers and how it may inform their decision-making. The life cycle costs of a building should not be seen as someone else's problem a long time after the design team has been disbanded. It is the responsibility of everyone in the design team to deliver buildings that take account of issues of sustainability and their operational costs as these are a major concern to local communities and society in general.

Fact file
There are a number of definitions of facilities management but the following one probably best encapsulates the term: 'The process by which an organisation delivers and sustains a quality working environment and delivers quality support services to meet the organisations objectives at best cost' (The Royal Institution of Chartered Surveyors 1999).

Practitioners and students of design cost management are reminded that client and organisational and operational issues are part of a separate problem. They are essential prerequisites of all those participants involved in the design and construction process which need to be considered as early as possible when a client decides to commission additional building space, either existing, leased or new, as shown in Fig. 0.1.

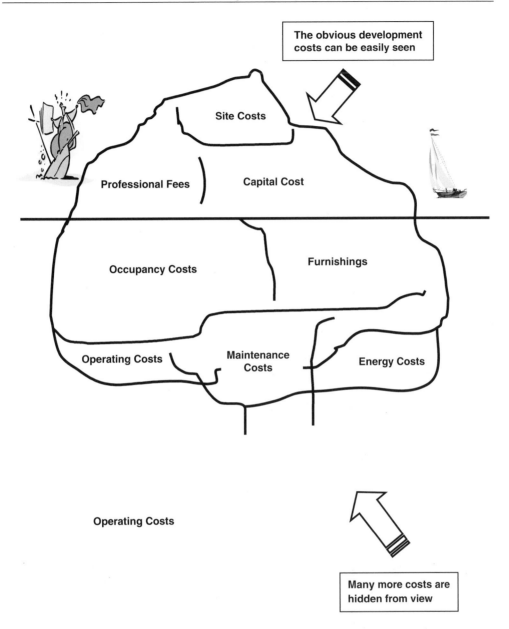

Figure 2.5   Components of life cycle costs. Source: Adapted from Flanagan and Norman (1983).

2.5   Summary

This chapter has defined design cost management and has traced its development since its inception in the 1950s by discussing the various influences that helped establish it as a recognised discipline. Reference has also been made to the important role that the BCIS has played in acting as custodians and organisers of information relating to buildings and their financial implications. An equally important role that the BCIS has played, is in helping to establish the academic base of design cost management and bringing rigour in its execution.

2.6   Reader reflections

- What do you understand by the term cost-planning infrastructure?
- The BCIS has been called a strategic cost information service. Why might this be the case in the context of the construction industry?
- How might the participants shown in Fig. 2.1 vary with, say, a design and build procurement arrangement?
- What do you understand by the term life cycle costs and how may capital cost influence such costs?

3 Design cost management and changing trends in construction procurement

3.1   Introduction

This chapter aims to:

- Introduce the reader to the context of design cost management
- Highlight the fragmentation that the construction procurement supply chain encourages
- Discuss how project objectives of time, cost and quality are balanced by the various procurement strategies
- Review how the various procurement strategies in use in the UK have evolved.

3.2   Context

Design cost management evolved and matured within a competitive lump sum tendering environment, established in the nineteenth century, and still remains the most predominant means of tendering, although its use has been in decline over the last 20 years or so. The success of this approach, together with the reasons why, have already been described.

However, as already mentioned, this panacea has, over many years, been severely criticised and various alternatives proffered. There is no doubt these criticisms have gathered more pace and volume over recent years. So what are the problems with such a simple and effective device?

They can be summarised as follows:

- The process of lump sum tendering, especially open competition, in which as many contractors as may be interested can bid, is wasteful of the industry's resources, especially in times of boom, as the considerable cost of preparing tenders, both successful and unsuccessful, is passed, ultimately, on to the industry at large, in the cost that it charges for its product: the built environment. There have been various government reports produced over many years, from as early as 1944 through to the present day (Simon 1944, Emmerson 1962, Banwell 1964, Latham 1994, Egan 1998), advocating the use of limited competition when only serious and capable contractors are invited to bid, and the number invited is limited, depending on the size, complexity and value of the project.

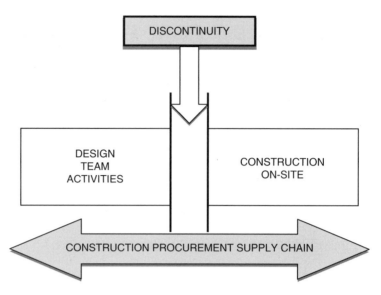

Figure 3.1 Discontinuity in the construction procurement supply chain.

- This process has led to a unique and undesirable situation that does not prevail in any other industry, i.e. the separation of the processes of design and construction, both in terms of responsibility and chronology as illustrated in Fig. 3.1.

Such a state of affairs has led to the following difficulties:

- A mistrust between the client and his or her representatives, the contractor and the subcontractors, both domestic and nominated. Perusal of the technical press reveals the very adversarial nature of the construction industry and, consequently, the number of referrals to litigation and arbitration. Clearly, such difficulties are not in the best interests of any of the participants in the industry, be they passive or active, as considerable intellectual energy, time and money is invested in attempts to resolve these difficulties. In recent times there have been attempts to mitigate these problems by introducing alternative means of dispute resolution such as adjudication, mediation and reconciliation. Evidence of these recommendations can be found in a number of recent reports such as those prepared by Latham (1994) and Egan (1998). Additionally the Construction Act (1996), the full title of which is the Housing Grants, Construction and Regeneration Act, which became law in 1998, has incorporated many of these changes. This state of affairs is further exacerbated by the fact that often the client and his or her representatives, the main contractor and the various subcontractors and suppliers, may never have worked together, hardly a recipe for establishing harmonious and cooperative working conditions. Attempts have been made to minimise this situation by the preparation of selective lists of potential tenderers, who are known to the client and his or her team. Such mistrust is also reinforced by the approach to the education of the future members of the design and construction teams. Unfortunately the various professional bodies have tended to create exclusivity in their approach. Hence, new members tend to be educated without having sufficient awareness or understanding of the other members of the team, their duties, functions and *modus operandi*. Again, this has been recognised in various significant reports (Andrews and Derbyshire 1993, Latham 1994) in which the problem has been identified and possible solutions recommended but, as yet, there has been little uptake in resolving this issue of paramount significance. This separation of designers and constructors means that design solutions do not always create the most efficient and effective solutions in terms of their on-site construction, as they often lack a practical approach, or 'buildability'. This characteristic will be considered later, when discussing the application of design cost management in the case study. Additionally, as a result of the separation of design and construction, it is usually the situation that the period from inception to hand over of the project will be longer, since there is no opportunity to overlap the processes of design and construction. This limitation can be avoided in other procurement arrangements as will be highlighted later.

■ Risk avoidance, through risk transfer to the main contractor, at first glance seems an attractive benefit to the client, and in many instances this is so. However, there are many situations where such risk avoidance may not always be desirable and, in fact, may not prove to avoid the risks at all. For example, the risk of time delay is very difficult to avoid in most lump sum contracts, despite mechanisms in the forms of contracts purporting to set up procedures to protect the client. A further example may be that, in a particularly complex project, the proposed design solution might prove unworkable and result in expensive and time-consuming delays and changes. Recent technical press articles demonstrate such difficulties, perhaps one of the most spectacular topical example being the Millennium Bridge across the Thames and its 'wobble'.

Additionally, risk avoidance may not necessarily be so desirable and might be better born by the client. For example, a procurement strategy might be not to allow for fluctuations in the prices submitted by the contractor at the tender stage. Such an approach, especially in times of high inflation and a booming economy, may well lead to a more expensive ultimate solution than if the client had agreed at the tender stage, to reimburse the contractor for any future increases due to fluctuations in prices.

These difficulties are further exacerbated because of fragmentation and lack of transparency in the project information produced and used by the various participants, as a basis for contractor selection. This information provides the basis for the management of the on-site processes of construction, which leads to breakdown in communications, resulting in mistrust, confusion and mistakes. These problems, created by the separation of design and construction, have been highlighted for many years, the most telling report being that by Higgin and Jessop (1965), which focused on problems created by fragmentation and opacity caused by the lump sum tendering process. Fragmentation and opacity have resulted from the production of tender documentation, which is poorly coordinated and not suitable for the generation of an integrated process model to guide management in carrying out construction on site.

Paradoxically, at the heart of these difficulties is the bill of quantities, which, as highlighted in Chapter 1, provides the financial management for building projects, where they are used as a basis for tendering, as they facilitate the various benefits advocated in Chapter 1, including the production of cost analyses for use in design cost management. There is poor coordination between the project information, produced by the design team, in the form of drawings, specifications and the descriptions and quantities of the work required contained in the bills of quantities. As a result, it can be difficult to find complementary information in the drawings, specifications or bills and often there are contradictions, conflicts and unnecessary duplication of the same information, causing confusion, time-consuming information searches and a high incidence of errors.

Considerable investment was put into the resolution of this difficulty, initiated by studies undertaken by the Building Research Establishment (BRE), who carried out various studies between 1978 and 1983 to identify the nature of the problem and to recommend moves towards its resolution (CPI 1987a). As a result the Co-ordinating Committee for Project Information (CCPI) was established in 1979 (CPI 1987a), which in itself was a major step forward in bringing the various professional domains together. Until this point there had never been professional cooperation in considering the problem of project documentation, information and its coordination between the architects (RIBA), the quantity surveyors (RICS), the engineers (Association of Consulting Engineers—ACE) and the builders (Building Employers Confederation—BEC). As a result of their endeavours, they produced a set of codes of practice, recommending how drawings and specifications, together with descriptions and quantities of measured work, should be arranged and presented, to minimise confusion and misunderstanding (CPI 1987b, c, d, e, f, 1998). The recommendations were based on a common arrangement of work sections, which was at the heart of the approach. The format of the Standard Method of Measurement (SMM) was also designed in accordance with the common arrangement of work sections, thus encouraging quantity surveyors to produce their bills of quantities compliant with the coordinated project information (CPI) recommendations.

Unfortunately, due to inertia and reluctance by the professions concerned, the uptake of CPI has been somewhat limited, despite the recommendation, in the Latham report (1994), that it should form the basis of the project documentation and that a similar development should be established in the civil engineering industry. Perhaps more success would have been achieved had it been made a mandatory requirement for the arrangement and structuring of project documentation.

However, there are some encouraging signs, in that Uniclass (1997) has now been established, which sets out to provide a means of classifying construction information, including both building and civil engineering, which embraces, not only the physical entity of projects, but also the use of the spaces within them.

Fact file

It is the view of the authors that, with the rapid advances in information technology, the need for generic information presentations, as is currently the case, will disappear, and be replaced by more dynamic information presentations for particular purposes at specific times. Such approaches will necessitate effective information organisation, provided by means of classifications such as that provided by Uniclass, coupled with a more resource-based approach to reflect, with greater accuracy, the process of construction. Chapter 11 discusses in more detail how such organisation might be achieved through the use of classification systems such as CPI and Uniclass and how more effective design cost management might be brought about.

3.3   Alternative procurement strategies

Unfortunately, there is no one perfect procurement strategy, as evidenced by our discussions pertaining to lump sum tendering. Each individual approach will be more or less effective in a given circumstance. The obvious solution (not easily achieved) is to attempt to select a procurement strategy, in which, in a given situation, the benefits outweigh the disadvantages.

It is perhaps worth reviewing, at this stage, precisely what is required of a procurement strategy. Put very simply, we are trying to optimise the triangle of cost, time and quality for the benefit of the client (Fig. 3.2). Unfortunately, these three expectations are usually competing with each other. For example, a very short construction time may well lead to higher costs. Low costs will usually lead to lower quality so, ultimately, a compromise must be sought which gives the best balance for a particular circumstance, as interpreted by the design team, acting on behalf of the client.

The major factors that influence the choice of a suitable procurement method are many and varied. In order to proceed with the identification of the most appropriate procurement strategy to achieve the optimal balance of time, cost and quality, the following criteria should be considered:

■ *The economic use of construction resources.* In times of boom the use of competition may be inappropriate, as there will be more work available than contractors able to carry out the work, and, therefore, the alternative of direct negotiation may be more sensible. Obviously the reverse scenario is equally true, except if there is only a particular specialist contractor able to carry out the work, then open, or even select competition will be inappropriate.
■ *The need for the contractor to contribute to the design as well as the construction process.* We have already mentioned the issue of buildability and fast tracking the design and construction stages by overlapping design and construction. However, where the project is simple in nature, highly predictable and time for hand over is not a major issue, the contractor's involvement in the design is not necessary or desirable. A typical example might be the building of local authority primary schools, when the authority is very familiar with the design, and time may not be an issue.
■ *The incentive to make production cost savings and their subsequent control.* Clearly, the best way to ensure the contracting firm does not use its resources inefficiently

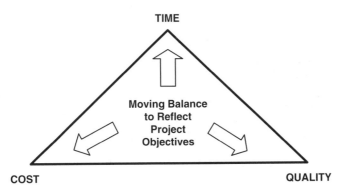

Figure 3.2   Project objectives (Rowlinson and McDermott, 1999).

and ineffectively, is to tie it into a commitment to carry out the given work for a fixed sum of money, as in a lump sum tender, or a unit rate approach. In such arrangements, the contractor cannot vary the price, but might argue about the quantity of materials used and the time required to utilise labour, plant and supervision. However, for reasons discussed earlier, in terms of achieving buildability and time savings, it may be beneficial to use a procurement strategy which allows for resources to be determined, together with their cost implications, at the time of their utilisation on site. These issues will be considered later. A major concern with such strategies is their subsequent control, to ensure the contractor is acting as diligently as possible in the carrying out of the works. Unfortunately, expensive and time-consuming monitoring is required, on site, by the client, to ensure that such compliance is met. To demonstrate the difficulty, the contractor may have installed an expensive tower crane on site, on a cost reimbursement basis, which is clearly being under utilised. A further example may be the possibility of more labour being on site than is actually required. It is also important to ensure that materials are not being paid for in a way which includes excessive, and perhaps bogus, waste allowances.

■ *Continuity of work.* The benefit gained from one contractor carrying out a series of similar kinds of projects, such as libraries, houses or schools, can be considerable, as the benefit of the learning curve should bring about improvements in efficiency, as well as improving the design and construction solutions, for the good of all concerned. Such arrangements are usually termed serial (or continuation) contracts, where the tendering contractors know, at the time of tendering, the extent of the series of the projects involved. The main point is that the projects in the series must be similar in nature to gain the benefit of organisational learning. Difficulties can arise when the contracting organisation becomes over stretched, and it may get into difficulties.

For regular and routine maintenance work, term contracts are often arranged, where the appointed contractor carries out the maintenance work, such as painting and decorating, on a large factory, educational institution or housing estate, over a predetermined time period.

■ *Risk and the assessment of who should bear it.* As has already been discussed, risk, and whether it should be avoided or transferred, is very much at the centre of any procurement strategy, and much has already been said about it, within the context of competitive lump sum tendering. Risk is not just financial but is also about whether the right quality can be achieved and whether or not the project can be completed on time. Nonetheless, in most studies, risk is assessed in terms of cost, which can be used as a common denominator to measure and compare all risks.

Risk can be broadly classified as follows (Latham 1994):

(1) Fundamental: war damage, nuclear pollution, supersonic bangs
(2) Pure: fire damage, storm
(3) Particular: collapse, subsidence, vibration, removal of support
(4) Speculative risks: ground conditions, inflation, weather, shortages and taxes.

All of these risks can be transferred by the client to the contractor. However, one needs to ask, 'Is such a strategy necessarily in the best interests of the client?'

The carrying of risk will obviously carry a cost premium and the more uncertainty the higher that premium will be. The judgement to be made, therefore, is whether the risk is better born by the client, as he/she may be better able to manage it. However, it is likely that the carrying of the risk by the client will prove to be a lower cost than that included in the contractor's premium.

It is also worth stressing that, what might be seen as an appropriate strategy concerning risk in, for example, the UK, may be seen as entirely inappropriate in another country, due to cultural differences. For example, in the People's Republic of China, the concept of competition and transferring risk to an independent contractor is an unfamiliar concept, due to their long history of a non-capitalist culture. Conversely, the approach observed by one of the authors whilst working in Saudi Arabia was to transfer as much risk as possible to the contractor. To give a specific example, the ground conditions were left entirely up to the contractor to take account of in his or her tender bid independent of ground water or the nature of the soil itself. Here, in the UK, we tend to recognise that the risk for uncertainty, in terms of ground water and soil conditions, such as unexpected rock, should be partly born by the client. In other words the message is, 'Don't assume your particular experience is necessarily the correct one.'

Fact file

It is not the purpose of this book to discuss risk in any depth, except to stress the point that risk avoidance and transfer is at the heart of most procurement strategies, and their consideration is pivotal in arriving at a decision as to how to proceed to select a contractor. There are numerous references on the subject (Hayes *et al.* 1986, Raftery 1994).

As previously discussed, lump sum competitive tendering was the predominant means of contractor selection, since its initiation at the beginning of the nineteenth century and remains prominent today. As of necessity, a variant of lump sum tendering is the use of bills of approximate quantities, where uncertainty in the amount of, or indeed the nature of the work required, is recognised at the design stage and, in essence, a re-measurement contract is established. This simply means that, as the work is actually carried out on site, it is subject to re-measurement to take account of any changes. Such contracts invariably form the basis of civil engineering contracts, where due to the uncertainty of the work, because of its 'one off' nature, together with often-uncertain ground works, there is general recognition, by all parties, that the work will usually be subject to re-measurement, as the work proceeds on site.

Where the project under consideration is either small in value and/or simple in terms of its construction, a lump sum contract will be deployed, but no bill of quantities will be provided, only a specification of the materials and workmanship to be deployed, together with a statement of the work to be carried out. Usually the specification is prescriptive in nature, in that it states, precisely and unambiguously what material and workmanship are required. Therefore, it does not encourage innovation, creativity or alternative ways of providing the required level of performance and function. However, it is also possible to specify, in performance terms, either for the whole project (design and build) or parts of it. This has the benefit of encouraging innovation, as the contractors involved are persuaded to develop a solution that meets the specified performance requirements. High level services, such as air conditioning, lift installations, etc. are particularly suitable for specification by performance, due to the many alternative available solutions.

Where bills of quantities are not used, the benefit is that it saves the client the quantity surveyor's fees (1–3%) involved in their production. The disadvantages are the loss of all the benefits stated in Chapter 1, which, if changes are required in the design solution after the selection of the contractor, could make the resolution of the financial management of the contract, at best, difficult and, at worst, contentious.

As a result of client dissatisfaction with the so called traditional procurement approaches, a number of alternatives have evolved, which, in the main, are concerned with providing the client with a better service through a more effective interface between design and construction. These changes allow an improvement in the buildability of projects, through the constructor having a direct input into the development of the design solution. This gains the benefit of allowing the design and construction process to overlap, thus saving on the overall time required for design and construction. Additionally, these variants allow risk avoidance and transfer to be distributed to and carried by (depending on the particular needs of the client and the project) those best able to manage such risks. A variant of the traditional lump sum approach was the development of two-stage tendering, which attempted to gain the aforementioned benefits, whilst retaining the benefit of competition. Here the contractor is appointed at the early stage of the project's design, based on financial criteria from previous projects and stated unit rates pertaining to the work envisaged such as concrete work, brickwork, steelwork and other significant trades and elements. This is carried out in competition in order to select the contractor who is able to carry out the work for the least cost. This cost information is then used to reimburse the contractor as the work proceeds. Of course, such a procedure carries more risk than a lump sum approach, as there is no initial tender figure agreed to by the successful contractor.

The first major departure from the traditional methods, which came into prominence in the 1970s and 1980s, was management contracting. These were in the form of management contracts, which were partially replaced by a further derivative, construction management contracts. These two forms of management contracting accounted for just over 20% of the value of construction work at their peak in the 1980s although in the 1990s their use declined to a market share of around 10% (Royal Institution of Chartered Surveyors 1994).

Management contracting was seen as a possible panacea to construction conflict and to reducing the overall period from inception to hand over. They were intended to provide a spirit of cooperation between the various participants to the contract, by allowing the contractor's expertise to be involved in the development of the design and allow overlapping of the design and construction stage.

In the first variant of a management contract, the client appoints a management contractor to manage the actual construction of the building for a fee. The management contractor acts like any other member of the design team, in that there is no entrepreneurial conflict and this new design team member can focus their expertise entirely for the benefit of the client, rather than for profit maximisation. The system works by the splitting down of the construction work into a series of discrete packages. Interested subcontractors are then invited, usually on limited financial competition, to bid for the particular package and if successful, carry out the work in the particular package, under the control of the management contractor. Construction management is a similar approach, except that the various subcontractors are contracted to the client rather than to the management contractor, as in a management contract. The theory is that the construction manager carries no risk and therefore has sole allegiance to the client and the project.

There is no doubt that management contracting has enjoyed a degree of success, especially where long-term relationships have been established and the client has a need for a regular building programme. One of the best examples is that of the major UK retailer, Marks and Spencer and the contracting organisation, Bovis, who have acted under construction management arrangements, over many years.

Perhaps one of the reasons for the failure of management contracting to gain a larger share in the market was because of the often acrimonious relationships between management contractors and the subcontractors. Often the contracts that were used by the management contractors with the subcontractors were heavily in the contractor's favour, and many subcontractors were considerably disadvantaged. They became unhappy at entering into such relationships. Clearly, construction management was seen as a way of alleviating this difficulty but client's perceptions of management contracting, in many cases, persuaded them that such an approach was not necessarily the most appropriate for their needs. A further reason for its possible decline is that most of the risk (financial, time and quality) is borne by the client and very little can be transferred to the other parties to the contract. This difficulty is highlighted by the fact that there is no predetermined contract sum.

As a result of these misgivings, the use of the design and build method of contracting increased dramatically in the 1990s going from a 10% share during the 1980s up to a 35% share of the construction procurement market, with management contracting declining to a 10% share (Royal Institution of Chartered Surveyors 2000).

The major advantages of design and build are that all the risks, both financial and period for completion, are transferred to the design and build contractor, together with the fact that the client is only dealing with one organisation, so eliminating the complexities and frustrations of dealing with a range of separate organisations. The approach also overcomes the problem of the separation of design and construction, as in the traditional approach, so saving overall time and allowing the design to reflect improved buildability in the construction solution, as obviously the design and build contractor can ensure that the proposed design solution reflects its own particular expertise and resource availability, in terms of developing the construction solution.

There is considerable debate as to the merits and demerits of the approach and it is fair to say that many design professionals were unhappy about the approach, not least because there was a reduction in their own influence and independence. However, a valid view put forward, which undoubtedly design and build suffered from in the 1980s, was that the quality of the final building was often inferior and the role of design was devalued, as the design and build contractors 'shaped' the design to suit their particular methods of construction. Consequently, the design may not have aspired, nor achieved the best performance and functional requirements for the client.

Measures have been taken to overcome this very real problem by introducing the concept of novation. Here, an overall design is developed by the client's own designer and that designer is then novated, or transferred to the design and build contractor, to ensure that the client's interests are best served. This approach, although sound in theory, has not always been met with enthusiasm by either the design and build contractors, architects or clients. This 'arranged' marriage can result in strained relationships.

There is a range of other procurement strategies that have a relatively minor role in the procurement of buildings, which, for the sake of completeness, are mentioned here.

One particular group of procurement strategies in this category are those known as target cost contracts. These again attempt to create a greater degree of harmony within the design and construction team, thus creating a spirit of cooperation, together with improvements in buildability and overlapping the design and construction stages, whilst still retaining an independent designer. This approach achieves similar benefits to management contracting as previously discussed, except there is slightly less risk to the client in terms of certainty of price, since, at the early stages of the design stage, an estimate of cost or target is established, which is made up of a prime cost plus an agreed allowance for profit. The actual prime costs are then recorded as the work proceeds, together with the agreed overheads, including profit. Any savings or additions are then shared, usually on a pre-arranged percentage allocation, thus dividing the financial risk between the client and the contractor. Such contracts are often used in projects involving a high degree of risk, such as marine works and tunnelling. Another little used approach, is the cost reimbursement contract. Here the client usually negotiates with the contractor to carry out the work for the prime cost, together with a separate fee for general overheads and profit. Clearly such an approach carries maximum risk to the client, as there is no estimate of cost to which the contractor is committed. In order to make the contractor inclined to be more efficient and effective with the use of the resources, the fee may be fixed, rather than assessed on a percentage basis.

Such arrangements are often used in emergencies, when arguing about financial implications may be some way down the list of the client's priorities if, for example, the roof has blown off or a basement has flooded!

This ends a brief review of the various procurement strategies that have evolved and are in use in the UK. As highlighted in Chapter 1, for a more detailed and comprehensive treatise of all the various procurement strategies discussed, the reader is referred to the literature (Aqua Group 1990, Masterman 1992, Morton and Jaggar 1995, Rowlinson and McDermott 1999). Before discussing the role of design cost management within the procurement process, Chapter 4 highlights the latest thinking and recommendations, in order to gain some perceptions of future trends in construction procurement and their likely impact on the design cost management process.

It is a matter of fact that, unfortunately, the construction industry remains, in many cases, unable to perform, in terms of delivering projects of the right quality at the right price and at the right time. Stark evidence of such under performance is highlighted in Fig. 3.3, where failure to deliver buildings on time and within budget occurs in 70% or more projects.

73% were over budget

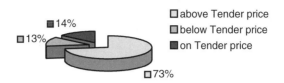

70% were delivered late

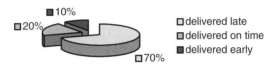

Figure 3.3 Performance of the UK construction industry. Source: Benchmarking the Government Client (1999).

3.4   Reader reflections

- Why do you think the UK construction industry is sometimes considered as adversarial in nature?
- What initiatives have the UK government put in place to reduce the incidence of disputes between the parties involved in a construction project?
- How does the bill of quantities introduce opacity into the relationship between the client and the contractor?
- What are the benefits of having coordinated project information to the various parties involved in design and tendering?
- How are risks of design and construction allocated by the procurement strategies available to clients?

4 Future directions in construction procurement

4.1 Introduction

This chapter aims to:

- Highlight the recently published government commissioned reports that recommend significant action
- Discuss the drivers to achieve better construction
- Identify the key processes required to achieve change
- Suggest the need for a comprehensive information system that can support design cost management irrespective of procurement arrangement.

A number of major reports have recently been commissioned at government level addressing the problems facing the construction industry and what measures are required to bring about improvement. Construction procurement strategies are the main focus of all these reports.

4.2 Pressures for change

The first major report was produced by Sir Michael Latham in July 1994 (Latham 1994) and, in fact, echoed many of the criticisms in the earlier reports highlighted in Chapter 3, which had largely been ignored. It sought the views of contractors and key private and public sector clients and set out a clear action plan, which, if implemented, would potentially make efficiency savings in the order of 30% in total construction costs over 5 years. An additional key recommendation was that the government itself should commit to becoming a best practice client.

Amongst the many recommendations the following were the most significant (Latham 1994):

- Legislative procedures to simplify the resolution of disputes through mediation, reconciliation and adjudication as discussed earlier, together with ensuring prompt payment for work carried out. Much of this was implemented through the Housing Grants, Construction and Regeneration Act introduced in 1996 (Construction Act 1996).
- The establishment of a single organisation to bring together all sections of the industry and clients. This resulted in the establishment of the Construction Industry Board. It also led to the establishment of the Construction Clients Forum, a separate organisation, representing clients.

■ The publication of a number of guides, checklists and codes of best practice concerning various aspects of the procurement, design and construction processes.

Following on from the Latham report was 'Rethinking Construction' by Sir John Egan (1998) that set out a number of key drivers and processes which needed to be put in place to secure significant improvements in construction performance. The five key drivers identified to achieve better construction are (Egan 1998):

(1) Committed leadership
(2) Focus on the customer
(3) Integration of the process and the team around the project
(4) A quality driven agenda
(5) Commitment to people.

The four key processes needed to achieve change are (Egan 1998):

(1) Partnering the supply chain by establishing long-term relationships based on continuous improvement with a supply chain
(2) A sustained programme of improvements for the production and delivery of components
(3) Integration and focus on the construction process and on meeting the needs of the end user
(4) Elimination of waste in the construction process.

The report went on to identify seven annual targets capable of being achieved to improve the performance of construction projects (Egan 1998):

■ Reductions in:
    (1) capital costs of 10%
    (2) construction time by 10%
    (3) construction defects by 20%
    (4) accidents by 20%
■ together with increases in:
    (5) predictability of projected cost and time estimates by 10%
    (6) productivity by 10%
    (7) turnover and profits by 10%.

In order to provide momentum for the recommendations from these and other reports the Government has, through the Department of the Environment, Transport and the Regions (Comptroller and Auditor General 2001) launched a number of initiatives as follows:

■ *Construction line.* This was launched in 1998 as a public–private partnership to provide a qualification service as to the suitability of contractors and consultants to undertake design and construction work.
■ *Movement for innovation.* This was also established in 1998 to promote innovation in construction and share good practice. A particular feature is the encouragement of contractors and clients to put forward examples of good practice by means

of demonstration projects. At the time of writing there are some 170 or so such examples.

■ *Construction best practice programme.* This programme was established in 1998 with the aim of raising awareness across the industry of the need for change and to identify good practice, together with its dissemination to the industry.
■ *Housing forum.* This was launched in 1998 to take forward improvement initiatives directed at the housing sector.
■ *Local government task force.* This was launched in 1999 to promote the principles of good practice set out in Egan's report (Egan 1998).

As a result of all these various initiatives and pressures, six essential requirements for all construction projects, have been proposed for the effective procurement and management of construction (Comptroller and Auditor General 2001):

(1) *Contractor selection.* Contractors should be selected on the basis of achieving long term sustainable value for money and not just lowest price.
(2) *Integrated design and construction.* Construction design should not be a separate process but be integrated with the whole construction process so that the design team can take more responsibility for the implications of their design, including cost, quality, buildability and the health and safety of those required to construct, renovate, refurbish, maintain and demolish buildings.
(3) *Better planning.* Sufficient time should be given to planning before commencing construction. This involves:
  (a) getting the construction sequence right
  (b) the assessment of risk and its management
  (c) carrying out value management.
(4) *Project management.* The establishment of reliable project management. The characteristics of good management are:
  (a) comprehensive understanding
  (b) detailed knowledge of risks
  (c) regular monitoring
  (d) effective communication.
(5) *Benchmarking.* The performance of construction projects should be measured to assess whether cost, time and quality requirements are being met and to learn to disseminate lessons for future projects.
(6) *Fair price and better value.* Contractors should be remunerated in a way that encourages them to deliver good quality construction on time and to budget.

■ *Information system.* A further essential requirement to bring about better construction performance which has already been highlighted in Chapter 3 is the development of a comprehensive information system, underpinned by information technology, which facilitates the accurate and rapid provision, manipulation and assembly of the specific information needs of all those concerned in the design, construction and operation of construction projects over their total life cycle, as a basis for their effective and efficient management.

Figure 4.1 has been developed from the Comptroller and Auditor General report (2001), which sets out what is needed if better construction performance is to be achieved.

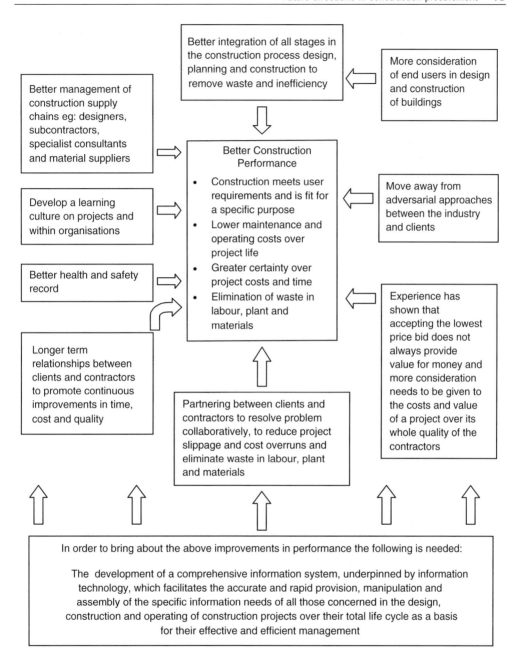

Figure 4.1   Steps necessary to bring about better construction performance.

Perhaps the main movement for change that comes out of all these reports and initiatives is the move towards greater harmony and spirit of cooperation between the various parties involved in the procurement process and the elimination of the adversarial culture with all the difficulties, tensions and conflicts, that so often prevail within the construction industry.

Partnering is, without doubt, seen as the way to achieving many of the improvements reported in this chapter, if we are to establish a modern, forward thinking, innovative construction industry. There are many useful publications explaining in detail the approach to partnering including its implementation, together with its advantages and disadvantages (Bennett and Jeyes 1995, Hellard 1995, Construction Industry Institute Australia 1996, HM Treasury 2000) but it is worth commenting here on some of its key aspects.

Partnering is based on encouraging better relationships between all the parties and relies on the establishment of trust. It is a management tool intended to promote more cooperative working between the various contracting parties involved in a construction project. The primary objective is to establish the shared goal of completing the work needed in a cost effective and timely manner to the mutual benefit of all concerned.

Two major forms have been identified:

(1) *Project partnering*. This is where the parties work on a single project.
(2) *Strategic partnering*. This is where the parties work on a series of construction projects to encourage continuous improvement.

Protagonists of partnering (Bennett and Jeyes 1995, Hellard 1995, Construction Industry Institute Australia 1996, HM Treasury 2000) have identified the following specific benefits from clients and contractors working together which help in the achievement of better value:

■ Reduction of the need for time consuming and expensive design changes
■ Reduction of expensive disputes leading towards costly and lengthy litigation
■ Integration of the supply chain
■ Replication of good practice from earlier projects
■ Encouraging the contractor to contribute to developing more efficient and effective design and construction solutions
■ Improvement of project performance during its lifetime
■ Development of more integrated and shared information systems providing greater transparency and specificity for the various parties involved.

Whether or not partnering will prove to be the panacea that its various supporters claim, only time will tell. It is perhaps worth pointing out that such approaches are not new and similar arrangements and strategies have been deployed under various pretexts over many years (serial contracting, for instance) with limited success. Issues of probity in public tendering tend to work against any procurement strategy that wishes to develop a continuing relationship between client and contractor. Once the fragile bond of trust is breached the whole process can be undermined. Partnering may also inhibit competition and restrict the entry of new firms into the market. Ironically, some contractors may object to its universal adoption. Whilst a contractor may win a number of projects in a medium/long-term partnering relationship, a greater number of losing contractors miss out on these projects, which are now withdrawn from the market place. So, we have to be vigilant that competition is retained as much as it can with these partnering arrangements and to guard against complacency in innovation, cost reduction and operational efficiencies.

Despite the rhetoric, the culture of a large number of clients is still to sign contracts with the lowest tenderer, irrespective of the perceived benefits of more harmonious relationships. The prospect of paying less is just too tempting to many consumers—whether it is the purchasing of groceries or buildings!

Partnering is, without doubt, an excellent principle to follow but, changing the culture of the construction industry's participants to facilitate its full support, may prove difficult to achieve.

4.3   Future procurement proposals

The Comptroller and Auditor General report (2001) has recommended three procurement strategies for use with central government departments' construction projects within the context of partnering arrangements aimed at bringing about the various improvements sought in construction performance. These are:

(1) design and build
(2) prime contracting
(3) public–private partnerships.

Although these proposals may well be subject to review and change they nevertheless reflect the fact that the traditional long established approach to contractor selection within which design cost management evolved is likely to become used less and less as a basis for contractor selection in the future.

(1) *Design and build*. This approach has already been discussed in the Chapter 3. However, the point we wish to make here is that it is recommended, by the Office of Government Commerce, within a partnership arrangement as one of the major ways of procuring construction projects. So, traditional forms of procurement are no longer the only method that government can use to procure projects.
(2) *Prime contracting*. This is seen by the Office of Government Commerce as an extension to design and build in that the Prime Contractor is expected to have a well established relationship with a supply chain of reliable suppliers. The aim is to achieve increased quality and value for money because of better consistency of the finished work and greater use of standardisation. Again, partnering is a pivotal component of this approach.
(3) *Public–private partnerships*. Here a supplier is contracted to build a public facility or infrastructure such as a hospital, bridge or tunnel, or a prison and also to be responsible for running the service the particular facility is providing usually over a concession period of, for example, 30 years. In this way the risks associated with providing the service are transferred to those best able to manage them. Again, the principle of partnering, both at project and strategic level, is an integral requirement if a successful outcome is to be achieved.

Fact file The reader is recommended to consult the following reports and initiatives to gain further understanding of the improvements being sought, together with their implementation, within construction:

Constructing the Team 1994, Construction (Design and Management Regulations) 1994, Construction Industry Board 1995, Levine 1996, Efficiency Scrutiny into Construction Procurement by Government 1997, Government Construction Clients Panel 1997, Building Down Barriers 1998, The Pilot Benchmarking Study 1998, Construction Best Practice Programme 1998, Construction Task Force and its Report 'Rethinking Construction' 1998, Movement for Innovation 1998, Housing Forum 1998, Constructing the Best Government Client 1998, Achieving Excellence 1999, Commission for Architecture and the Built Environment 1999, Local Government Task Force 1999, Construction Industry Board Root and Branch Review 2000, Government Construction Client Panel 2000, Modernising Construction 2001.

4.4   Design cost management within construction procurement

We hope this review of procurement strategies in terms of current approaches and future directions provides essential background to our descriptions of cost management. In fact, one of the purposes of the review is to enable us to consider the future of design cost management within the changing approaches to construction procurement.

The purpose of this book is to explain, describe and demonstrate design cost management as it is currently developing. However, anybody involved in design cost management, either out of interest or to aid their professional life as construction economists, might be concerned as to its relevance and application within the context of the changing nature of the construction procurement process.

An important role of the bill of quantities is to provide the necessary cost planning data by means of elemental cost analyses. It can be seen from our discussion of contemporary trends in construction procurement that it is likely that there will be a decline in the production and use of bills of quantities as the use of lump sum tendering continues to decline. The changing trends in procurement, especially towards design and build arrangements are not likely to abate. Under such arrangements the bill of quantities in its current form is not an essential requirement for tendering and contractor selection, nor for the other various management functions identified earlier.

So what is the future of the discipline of design cost management? The first thing to note in all the reports and initiatives concerning improvements in the construction industry, is the ever increasing emphasis on the need for effective cost and time management in the design and construction process. In tandem with this is the need for greater recognition of the contractor's expertise in the development of the design, in order to use that skill and knowledge to improve the buildability of construction projects.

Within this scenario, the role of the construction economist is paramount. With the various tools and techniques available, the construction economist can model the likely implications of design upon construction, to provide a real opportunity to increase the objectivity, reliability and the integrity of the design cost management process. Currently, design cost management operates, mainly due to the failure of the bill of quantities to accurately model the resources needed in the construction process. Within this state of opacity, incorrect solutions can be generated and they are often sub-optimal solutions as the dynamics of the construction process are difficult to identify. Without doubt, moves towards greater transparency and sharing of information as provided by more integrated and harmonious construction procurement strategies, should provide the construction economist with the opportunity to move into a much more proactive and dynamic environment, than that currently in existence.

What will remain fundamental for the foreseeable future in terms of design cost management, is the facility to be able to adjust cost information relating to existing projects, in terms of time, quantity and quality. This is essential to produce a cost plan of the likely cost implications of the future project under consideration. In fact, the project used to describe and demonstrate the application of design cost management later in this book, is to be procured by means of design and build. However, to this current approach to design cost management the construction economist should be able to add a further important dimension; that of reflecting the *process* of construction as well as the *product* of construction. This should lead to greater reliability in our cost predictions, and equally important, will facilitate the establishment of more effective design and construction solutions, as the implications of the design on the construction solution will be much more explicit and better understood.

For this to happen, there has to be some parallel developments in information systems and their management in order to capture the opportunities of greater transparency offered by the developing of more open approaches to construction procurement.

The present time is an ideal opportunity to develop information systems which are project related rather than developed primarily for use by each specific contributor in the design and construction process. Such approaches in the past have led to fragmentation and poor portability, since they tend to only serve the needs of each particular interest. Through developments of classification systems such as SfB (CIB Working Commission W58-SfB Development Group 1973, CIB W74 Information Co-ordination for the Building Process 1986), CI/SfB (Ray-Jones and Clegg 1976), CPI (1987b) and latterly Uniclass (1997), which are further discussed in Chapter 12, it is possible for us to organise our information and there is now a real opportunity to utilise information technology such that a comprehensive information system can be developed. These developments can facilitate the accurate and rapid provision, manipulation and assembly of the specific information needs of all those concerned in the design, construction and operation of construction projects, over their total life cycle, as a basis for effective and efficient management.

4.5   Summary

In conclusion, there is a need for the construction economist to develop and establish an effective interface between the product stated design and the process stated construction solutions, by linking the current approaches and techniques of design cost management to those developed by construction managers. In this way they can thus capture functional product related information and the dynamics of the construction process, as manifested in the constructor's programme. More will be said about this later, but at this stage it is important to stress that the current skills, knowledge and techniques of design cost management remain paramount, and will continue to be so, within the construction economist's armoury, if we are to contribute to the effective resource management of construction projects.

We will attempt to integrate all this knowledge into a procurement and cost management approach for an electronics component production facility, taking the reader through the various steps in the process and commenting on the alternatives and rationale for making decisions.

4.6   Reader reflections

- What are the barriers to changing practice in the construction industry?
- Can the UK construction industry clients provide a continuity of work to ensure that changing practice takes root?
- What are the benefits of partnering?
- What are the potential pitfalls of partnering at a strategic and project level?
- What role would the independent construction economist play in a project partnering context?

5 Design cost management: models and data

5.1 Introduction

This chapter aims to:

- Introduce the reader to the concept of models and their use in construction management
- Discuss the limitations of the data that are used for building design cost models
- Suggest a brief taxonomy of cost models
- Highlight the use of the various models at different stages during the design process.

The process of design cost management is facilitated by the use of cost modelling as a means of seeking an optimal solution for the provision of a new construction project. The cost models are useful in that they are able to explain and account for the cost structure of construction projects. Needless to say, underpinning the application of modelling techniques is the need for accurate and reliable data.

The purpose of this chapter is to consider in greater depth the various modelling techniques available to the construction economist, and the data that drives them, in order to help manage the financial outcomes of construction projects. It is not the aim of this text to explain these techniques in great detail (as there are many texts which already do this), but to attempt to put these techniques into context within the design cost management process, both in terms of function and project knowledge processes.

5.2 Cost models

The purpose of models can be summarised as helping to provide the following information about the construction project system they represent:

- *The communication of facts about the system.* For example, we might use a model which will tell us the likely future cost of a primary school to be built in 2002 in Loughborough, Leicestershire.
- *The communication of ideas about the system.* For example, the model of the proposed space use within the building may give rise to a design review of what space to provide and its juxtaposition to other spaces.
- *The prediction of how the system will behave in certain circumstances.* For instance, the total costs of a project may change as a result of different procurement methods.

■ *The provision of insight into why the system behaves as it does.* For example, in observing the cost behaviour of our project model we might be able to assess how costs vary as the height of the building increases.

It is an obvious statement, but the reason models are of such benefit is because they can assist in predicting and explaining the behaviour of a real life situation. The prediction may have varying degrees of confidence, in the ways identified above so they can inform the decision makers as to the effects of making a particular decision. For instance, what would the effect be of raising income tax or VAT on a particular form of construction project or its content?

Models, as discussed earlier, can be simple reflections of reality such as a photograph, a painting or a three-dimensional model through to mathematical representations of a particular system, such as the economic system of a country or the structural performance of a framed building. Models that are most useful in performing the objectives stated above tend to have a mathematical base and usually need a refined degree of understanding to calibrate and manipulate them and fully comprehend the information they provide.

The most useful models for those involved in design cost management are cost models as the discipline of design economics tends to use financial information as a basis for giving advice and decision-making. The term cost modelling, although widely used perhaps needs some further elaboration. The term itself more accurately means price modelling, that is the price the constructor charges for constructing the project. This is because the disciplines of design economics and cost modelling have been developed by construction economists working for designers. Generally the financial data they have available reflects the constructors' prices (tender prices) for providing the completed construction work rather than the internal costs which such constructors incur in hiring and/or providing the basic resources needed to achieve the completed construction project. Even construction organisations tend to present their models in price terms where they are involved in the tendering process, attempting to identify the market price. However, to win the project the tenderer may have to submit a price just below the market price as judged by the other competitors. When the constructors themselves identify how the project is to be constructed in terms of the activities to be carried out, their sequence and what resources of labour, plant, supervision and materials are required, they are, in fact, using a true cost model or, to be more precise, a resource model. They ultimately convert this resource model into a cost model, which in turn after consideration by the board of the company, will be converted into a price model, usually by a financial addition, or tender assessment process. This may be large in times of booming conditions and conceivably as a reduction in times of a slump.

The limitations of cost models, as used by construction economists, have been highlighted earlier in that they give only indirect information on the real costs of construction: the resources needed. This is because the basic tool for constructing many such models is the tender breakdown contained in the priced bill of quantities. Although the bill of quantities is 'priced' by the constructor it is a reflection of the market price and is not necessarily a direct reflection of the resources needed, nor necessarily how efficiently they are being deployed. In fact, a tender may be considered a 'normative' model of planning, construction intent and cost structure based on the information available for the construction team to use as a guide (or a benchmark) for their actual work and operations. This is partly why design and build as a mode of construction procurement has become more widely deployed as the 'buildability' or efficiency of resource deployment can be better considered by the constructor responsible for the resource provision who can then establish a specific resource model for the particular project under consideration taking into account the particular skills, techniques and resources available to that construction organisation.

Comment has already been made earlier that the market price anticipated by the constructor may have little direct relationship to the resources needed to achieve the completed work. This was described in Chapter 1 with reference to the study carried out by Fines (1974) where bids for identical structures but with different functions (a barn and a theatre) had entirely different financial implications. Interestingly enough in the civil engineering world cost models based on price tend to have a much closer relationship to the resource implications and their subsequent costs. This is because civil engineering projects tend to be one-off and there is only a limited history of similar projects and therefore a somewhat limited history of the likely market price. Additionally, civil engineering projects are often fraught with uncertainty, due to the nature of the work involved, and so predictions of the likely financial implications of such projects are often at great variance with the final price paid to the constructor. To illustrate the point there are many construction projects involving schools, libraries and offices, but not too many pumping stations, railway tracks or motorways.

Work by Barnes in the 1970s (Barnes 1970–71) recognised these particular difficulties of trying to achieve equitable financial settlements of such work caused by the limitations of existing civil engineering cost models based on bills of quantities. Barnes introduced a much more realistic and accurate basis for the building of cost models by introducing a standard method of measurement which set out to take account of contemporary civil engineering methods, techniques and practices as well as reflecting the fact that much plant with low deployment flexibility, such as cranes, are used which have high installation and removal costs as well as high non-productive or time-related costs. A similar approach was also introduced into the standard methods used in the building industry (RICS and NFBTE 1978, CPI 1987c, 1998), but with more limited success than in the civil engineering environment. This was because the various standard forms of contract used in the building industry call for a planning programme to be provided by the contractor reflecting a model of the processes needed to be carried out on site to achieve the project. There are no guidelines or mandatory requirements set out as to its form, content, nor indeed its use. However, in civil engineering work, under their standard forms of contract, the nature and purpose of such programmes are much more clearly identified, for example, in clauses 13 and 14 of the ICE Conditions of Contract (Institution of Civil Engineers 1999) as well as in the Engineering and Construction Contract (1991).

As stated above the constructors' models, based on the processes of construction should, if prepared and used correctly, be an accurate prediction of the resources needed. Such models are generally in the form of network diagrams, precedence diagrams, bar charts, lines of balance or method statements as explained in various construction management textbooks (Cooke and Williams 1998,

Harris and McCaffer 1995, Fryer 1990). All of these, in varying degrees of detail, allow the activities to be identified, together with their relationship to each other and identify the resources needed. This brief description is an oversimplification of reality, as building and using such a model is a complex and highly sophisticated process. The constructor has to ensure that the scarce resources available are put to the best use. This is achieved by ensuring their optimal use over the period of time they are required and to ensure that there are no upper limit violations, which basically means that such activities needing such a resource allocation cannot be executed, unless more resources are made available. Figure 5.1 shows poor utilisation of labour resources together with an upper limit violation. Such models are invariably based on network analysis techniques and are underpinned by sophisticated computer aids to facilitate rapid iterations with various planning and construction scenarios in order to seek and present the optimal solution for the execution of the project.

A further inherent problem with our price-based cost models derived from the bills of quantities is that where lump sum tendering is deployed the price allocations are made often in a very short tendering period and, more importantly, before the detailed resource models are established by the constructor for the projects. An additional difficulty is that the part of the constructor's organisation concerned with pricing the bills of quantities (the estimating department) is concerned with establishing unit rates within the bill of quantities and the estimator's experiences are focused on the assessment and establishment of unit rates, rather than creating the resource model. Of course the inherent opacity within these

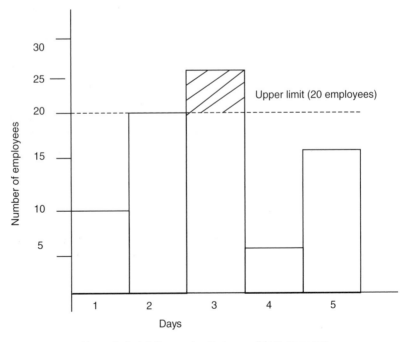

Upper limit violation on day 3—impossible to carry out

Figure 5.1   Resource levelling in planning.

price models can be of considerable advantage to the constructor in terms of improving cash flow profiles and to identify possible areas of financial leverage during the progress of the work.

A common practice is for the constructor to allocate distorted higher rates for work to be carried out early in the project. Ground works or structural concreting operations, for instance, may be the focus of distorted rates. Such a financial strategy is aimed at gaining, at the monthly valuations, an income greater than the costs that have been incurred up to that time. Obviously the model will be distorted in the opposite direction for the unit rates towards the end of the construction programme. Remember that the contract will have been tendered for in competition and, therefore, any unit rate adjustments have to be made within the overall tender figure. Such a strategy has advantages for all concerned as the client gets the work undertaken for a lower tender than otherwise would be the case and the constructor is able to more confidently carry out the project by minimising the financial risks associated with seeking funding from external sources.

However, there are some risks attached to such strategies from the clients' perspective, as should the constructor withdraw from site, become bankrupt or be forced into liquidation, then the client will have spent more money on the project than its actual value as well as receiving the building much later than originally anticipated. Additionally, where the constructor identifies a possible under measurement or omission of some work required in the project, then he or she will tend to record higher rates for such work, with the obvious expectation that when such omissions are identified and subsequently rectified, then it will be very much to the constructor's advantage. In fact a particular expertise of the quantity surveyor appointed under such lump sum contracts is to ensure that the various unit rates contained in the priced bills of quantities are accurate financial reflections of the finished work they represent. The ethical code of quantity surveyors acting in this capacity is to point out such variances in the pricing of these unit rates, whether they be to the constructor's advantage or the client's advantage, and to arrange for their rectification, as a prerequisite to commencing construction.

From the above discussions we can see that the models, usually available to the construction economist, are those based on the price the constructor charges for the descriptions (or specifications) of finished work as reflected within priced bills of quantities. We have also observed that these models contain a number of limitations that can make their applications complex where considerable care is needed to validate their robustness and at worst create erroneous or misleading conclusions.

5.3   A taxonomy of cost models

Before moving on to describe the various cost models that are made use of within the design cost management process it is worth stating here a typical taxonomy of models that are generally of use within construction:

- deductive
- inductive
- optimisation
- stochastic.

These four models are now discussed in turn.

5.3.1   *Deductive cost models*

These models are based on statistical techniques from which inferences, and/or trends, of anticipated behaviour can be established by examining statistically, large samples of data. Thus, these techniques cannot give absolute values with any confidence but may steer the construction economist towards a range of values within which the costs are likely to be found.

These cost models are particularly useful during the early stages of the design cost management process when little hard information in terms of detailed quantitative and qualitative information is available in terms of specifications, shape, height and other important design variables. Thus, these models seek to establish correlations, which can then be used to confirm a particular trend, either to confirm that the result was as expected, or to use the trend as a basis of prediction. One of the most widely used statistical applications is that of regression analysis where often cost is expressed in terms of cost per square metre of floor area (the dependent variable) against numbers of hospital beds, building height, etc. (the independent variable). A particular difficulty in the use of such models is that high correlations may be presumed to indicate causality, which in fact has nothing to do with the independent variable that is considered to be influencing the dependent variable. For example, increases in the cost per square metre of floor area might, from the regression analysis, be assumed to be due to the increasing height of the building. This could be an entirely correct assumption but it is just possible that such increases in cost per square metre are due to an entirely different or random reason, such as the different plan shapes of the various buildings making up the sample in the analysis.

Hence, these cost models must be used with caution to ensure predictions are made which are not outside the limits of the data. These models also have the limitation of being 'black box' (as distinct from 'glass box') in that it is not possible to see what is happening within the model and therefore the results have to be taken without explanation or justification, which may lead to erroneous interpretations, as highlighted above, but equally importantly throw little insight into why the model behaves as it does.

 Log on to the website to view a typical deductive cost model.

5.3.2   *Inductive cost models*

Inductive cost models tend to be based on exact algebraic expressions and therefore can be said to produce results which are causal rather than correlations as discussed above. Clearly these models are much more accurate and can be used with much more confidence to analyse and account for the cost of construction, and they can be used with greater reliability to predict the cost of construction at some time in the future. Perhaps the most widely known inductive cost model is that of the unit rate contained in the bill of quantities which, when priced by the constructor, becomes a cost model of that particular component of finished construction work.

Further examples of cost models are the elemental cost analyses that make up the BCIS, either in group or elemental form, together with the elemental cost plans that we produce as part of our design cost management.

Prior to the 1980s when low cost computing was unavailable attempts were made to build cost models based on simple algorithms which attempted to describe the geometry, function and specification of the proposed building. The best known of these models was the storey enclosure method, which is described and documented fully by Ferry *et al.* (1999).

Although these cost models based on causality are much more accurate than those based on statistics their main drawback is that they can only be applied when the project can be described in reasonable detail, more likely after the feasibility stage of the RIBA plan of work (RIBA 2000) in the design process.

Log on to the website to view a typical inductive cost model.

5.3.3   *Optimisation cost models*

The purpose of optimisation models is to seek the optimal solution or solutions against the given criteria, by searching the feasible solution space and finding the point or points which reflect the optimal solution (Fig. 5.2).

These models are difficult to construct and generally are most successful when optimising solutions to a given specific problem. Particular examples of their use are in optimising formation levels on undulating construction sites in terms of excavation and filling. This can be particularly complex, especially with regard to the differential cost implications of ground excavations compared with filling. Other typical applications are in energy calculations in order to optimise pipe sizes, window sizes, insulation, etc.

5.3.4   *Stochastic cost models*

Stochastic cost models are designed to take account of economic risk by using risk analysis techniques. The most generally applied technique is the Monte Carlo simulation which is based on simulating activities over time and thus a life history of the system to be studied. These modelling techniques are not readily acceptable, as traditional deterministic approaches tend to be more readily accepted rather than ranges to which confidence limits are attached.

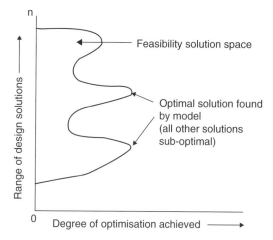

Figure 5.2   Cost optimisation models in design.

5.3.5 *Summing up*

There are a number of texts which explain in detail the nature, purpose and application of cost models as discussed above. In particular, the reader is referred to *Cost Modelling*, which gives a detailed treatise of the subject of cost modelling (Skitmore and Marston 1999).

All of these applications have developed extensively and will continue to develop, due to the rapid advances in low cost, universal computer software, such that now very powerful low cost and portable personal computers play an ever increasing role in the development and use of cost models. In addition to the developments in hardware, through the ease of use and flexibility of accompanying software, non-computer specialists are able to harness the technology without the need for specialist programming expertise, as was the case when computing first became available.

The construction industry at large, and construction economists specifically, have perhaps not taken up the full benefit that information technology can offer, especially with regard to developments in cost modelling. A particular problem is that construction economists, and the clients they serve, tend to look for deterministic cost predictions, based on single figure solutions and therefore most of our estimates are based on the establishment of single rate figures, whether they be budgets set at the feasibility stage or detailed unit rates established by constructors when pricing bills of quantities. Clearly such predictions are quite absurd for as with all the uncertainty attached in the design and construction of projects such accuracy is a fallacy. A more appropriate strategy would be to use more stochastically based cost models where ranges of cost prediction can be established and confidence limits assigned to such ranges across the various scenarios.

Work by Fortune and Lees (1996) has shown that the adoption of cost modelling techniques based on approaches, other than simple deterministic cost models, remain very much in the domain of a limited number of construction economists. As already pointed out, the nature of the data being used to populate these various models as used by the construction economist to derive single figure estimates further adds to the fallibility of such cost predictions. In fact, construction economists have developed techniques to hide such fallibility by the use of what are called price and design risks whereby the construction economist can attempt to steer the project as it develops towards the initial single figure budget by means of making adjustments to these contingency sums. Later in this book reference will be made as to what these techniques are and how they are used by the construction economist.

5.4 Modelling and design cost management

Having explained briefly the nature of design-based cost models we can now consider their form and use within the design cost management process as established within the context of lump sum tendering. We will use the well established RIBA plan of work (RIBA 2000) within which to place design cost management so that we can see what needs to be done, when it is to be done and how it is to be done, as we proceed through the design and construction process.

Figure 5.3 shows an alternative presentation of the RIBA plan of work shown in Chapter 2 (Fig. 2.1). This diagram show the sequential nature of the design

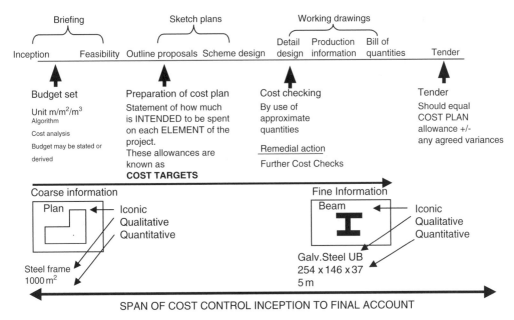

If fine information is available at FEASIBILITY stage cost planning process is redundant as cost targets can be set using the more reliable technique of approximate quantities

Figure 5.3 Sequence of design cost management: RIBA plan of work.

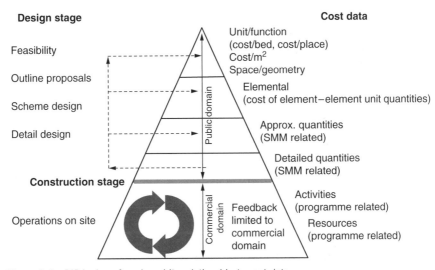

Figure 5.4 RIBA plan of work and its relationship to cost data.

cost management process, and the activities that need to be carried out as we move from inception through to carrying out work on site.

Figure 5.4 is an alternative presentation of the RIBA plan of work which has been adapted from Ferry *et al.* (1999). Its purpose is to show the nature of the data employed in the cost models at the various stages in the design cost management process.

It is noteworthy how all the models, apart from those used by the constructor shown at the bottom of the pyramid in Fig. 5.4, are all populated in one form or another with data derived from priced bills of quantities describing the price of construction. The contractor's model, at the bottom of the pyramid, makes use of resource information and therefore tends to reflect the cost of construction. A further point to note, as indicated in the diagram, is the cyclical nature of the contractor's model in that the cost data do not become available to the construction economist, but are internal and remain in the constructor's domain thus impeding the construction economist's ability to effectively manage construction costs. The diagram also indicates an interesting paradox in the design cost management process. Namely, as shown at the top of the pyramid, the fact that at the feasibility stage there is little project information available but an almost infinite amount of general information available in terms of possible solutions. The bottom of the pyramid shows that the amount of project information has increased to fully describe the project under consideration, even to knowing how many nails are required, which has been derived from the general information for incorporation into the project.

Of course, the project information itself ultimately becomes part of the general information, for use by others, for whatever purpose. This is an essential element of design cost management and it is only the information that is confidential to the provider (such as waste allowances on materials, labour outputs, etc.) that is the exception to the rule. In seeking the optimal solution it is important to remember that early on in the design cost management process all possible solutions are available, whereas later in the design, as we refine and decide upon the constraints, very few alternatives are available. For instance, once the major variables of plan shape, number of storeys, choice of frame have been made any non-conforming alternatives will have relatively little significance in terms of seeking the best solution. The reader is referred back to Chapter 1 to see an illustration of this important point. One can now begin to see why construction economists, armed with their cost models, play a pivotal role in helping find the best solution in terms of cost, quality and time.

It is perhaps worth pointing out that, if the client already knows precisely what is required at the commencement of the feasibility stage, then the concept of design cost management based on seeking the best solution through the use of modelling becomes less important as no major changes will be made to the initial proposal. To illustrate the point, if a local authority has an extensive school building programme, and all the schools are to meet the same performance requirements (i.e. 100-place primary schools) then one would expect the later schools to be built would be very similar, in terms of their definitive specification, to the earlier schools in the series. This would make the process of design cost management more focused and less broad ranging.

Thus, when we integrate design cost management into the RIBA plan of work we have to take into account three basic factors:

(1) the most appropriate cost models that can be used
(2) the sources and nature of data that are available
(3) how the chosen cost model will be used within the design cost management process.

5.5   Design cost modelling context

The first time we are likely to use a cost model is at the feasibility stage of the project which is the stage where we establish the technical and financial feasibility of the project. In other words, make decisions on what can be built and can we afford to build it? At this stage a range of costs will be established by the use of a number of deductive cost models to help establish a cost model for the project itself. Chapter 9 shows this process in detail with a project consisting of an electronic component production facility.

Essentially there are two approaches that can be made use of:

(1) The establishment of a cost range, by reference to a performance specification, expressed in terms of function, accompanied possibly by some indication of size together with its location and when the building is likely to be required.

> **Example**
>
> A 100-place primary school, to be built in a rural village in North Yorkshire, of 500 square metres gross floor area, in single-storey construction, and to be available for occupancy in 12 months time.

An alternative to this approach, which is often used at the feasibility stage, is to use cost models based on the cost per functional unit such as cost per pupil, patient, car space or cost per office worker, etc. Thus, if we know how many pupils the building is required to accommodate, then clearly from our cost model of the cost per pupil we can establish a cost range for the project.

> **Example**
>
> A further example might be to provide a prestigious office block of 10 000 square metres of usable floor area in a central London location, to be let speculatively. Note that when we price such a project an allowance must be made for the 'circulation' or non-rentable space (corridors, plant rooms, storage, lift lobbies, etc.) and in a prestigious project such as this one the percentage allowance may be in a range from 20 to 30% of the total gross floor area (GFA). We shall allow 20% for circulation space and the calculation of the GFA for the project is as follows:
>
> $$\text{Usable or rentable area} = \frac{10\,000}{0.80}\ m^2 = 12\,500\ m^2\,\text{GFA}$$
>
> $$\begin{aligned}\text{Cost range} &= 12\,500\,m^2 \times £1200\text{--}1800\ \text{per } m^2\,\text{GFA}\\ &= £15\text{--}22.5 \times 10^6\end{aligned}$$

(2) The second method is to use a Developer's Equation approach where an economic model is used to establish, from the rate of return (or yield), anticipated by the client, the amount of money that is available for the design and construction of the building, which can then be translated into a physical entity in terms of size, shape and specification. An example of this approach can be found in Chapter 9.

> The following equations show the contrasting, but related approaches:
>
> Size and specification, geometry, location and time = £ (Approach 1)
>
> £ = Size and specification, geometry, location and time (Approach 2)

There are a number of texts which demonstrate and explain the nature and application of the Developer's Equation, which the reader is referred to for further explanation (Enever and Isaac 1994, Baum and Crosby 1995, Darlow 1998, Isaac 1998, Isaac and Steley 2000).

The most generally available cost models are those which relate costs (the price the constructor charges) per square metre of gross floor area related to a particular function set against some predetermined point in time.

Example

For example, where the cost is £900 per square metre of gross floor area for 500 square metres for a primary school at March 2001. The use of such a model is obvious. The anticipated budget would be, as at March 2001:

$$£900 \times 500\,m^2 = £450\,000$$

Clearly this is a very simple approach and acts as a warning and an often true paradox in the application of design cost management:

The easier the use the less reliable the result.

5.6   Sources of cost data

As a result more sophisticated versions of the above have been developed together with other models, which primarily allow the construction economist to understand and take account of the implications of different locations and time. Construction economists tend to place more confidence in cost models, where the data that they make use of are very familiar to them. For this reason the larger organisations have their own cost models running with their own data. However, even the largest organisations find it difficult to keep wide-ranging information reflecting all the diversity of construction projects they are likely to be involved in. As a result attempts have been made to produce comprehensive published information, which is available to all those with an interest in the subject. These sources contain a variety of cost models together with 'best guess' data which range from a very coarse level of application, such as at the feasibility stage, right through to a fine level of application, for example at the Working Drawing stage of the RIBA plan of work (RIBA 2000). These publications are produced annually (Spons, Laxtons and Griffiths 2001) and the reader is recommended to review their structure and their contents. Practising construction economists will tend to resort to their use, either to confirm a particular opinion, or as a first source of information, when they are working in an area where they have little or no information of their own available.

5.6.1   *The BCIS*

One of the most widely used sources of information available to the construction economist today, is the BCIS, the history of which, has already been explained. We have placed a working model that utilises BCIS data on the website for the exclusive use of readers of this book.

Log on to the website to view the services and examples of data provided by the BCIS.

The strength of the BCIS is the comprehensive nature of its coverage, ranging over many different building types, reflecting many years of collection, together with a wide range of cost models from cost per square metre models through to detailed elemental cost analyses. The BCIS also has comprehensive models reflecting the implications of location and time on the financial outcomes of building projects. However, even the BCIS cannot reflect all the likely building types, especially in terms of size, morphology and specification. For example, there are very few football stadia to be found in the system, but numerous primary schools. Thus, if we are seeking information on football stadia the BCIS is not going to be very helpful but we will be able to find plentiful information relating to primary schools. We stressed in Chapter 1 'social opportunity costing' and the importance of seeking cost information on similar buildings or else we may arrive at entirely the wrong conclusions with potentially disastrous results. Again it is worth stressing that all the BCIS cost information is derived from priced bills of quantities with all the limitations that are inherent within these data as discussed in Chapter 2.

This leads us to another paradox that the construction economist needs to be aware of:

It is usually those building types where little information is available where such information would be of greatest benefit.

5.7   Reader reflections

■ Describe the uses of various models used during the design process. How do their forms vary with use?
■ What is the difference between construction price and construction cost?
■ What are the limitations of the cost models used by construction economists?
■ How might a resource-based cost model be developed based on a programme and method statement?
■ What are the differences between deductive and inductive cost models?

6 Design cost management: the feasibility stage

6.1 Introduction

This chapter aims to:

- Introduce the reader to cost information that can be used at the earliest stages of the design to develop the cost bracket
- Highlight the range of cost information and the statistical parameters that are used to assist in its interpretation
- Demonstrate how adjustments can be made to account for time
- Demonstrate how adjustments can be made for location
- Explain the basis for the various indices available to the construction economist
- Demonstrate how adjustments can be made for specification.

6.2 Estimating the cost bracket

Returning to our model of the design stages, the RIBA plan of work and the feasibility stage (Fig. 5.3), our cost models are, of necessity, coarse in their content, but if used with judgement based on experience, they can be used to set a realistic cost range for the particular project under consideration.

The most useful cost models available on the BCIS are statistical deductive models, which give a variety of cost information as shown below. Further examples are shown on the website:

Log on to the website to view statistical deductive models.

All the cost information contained in Figs 6.1–6.4 is brought to the same base date and location, by means of the BCIS themselves, using the Tender Price and Location Indices, respectively, as further explained below. This is necessary in order that we can compare like projects with similar time and equivalent locations when we are using the information.

Log on to the website to view average building price data.

The building function expressed in accordance with table 1 building functions, from the well established CI/SfB classification system, allows us to identify the particular building function at varying levels of detail by means of a description and a non-mnemonic code in the cases shown in Figs 6.1–6.4:

712.   Primary schools (GFA and functional area)
412.   General hospitals
275.5   Factories for electronics, computers, or the like.

Note, it is the function that is being classified and not the form (design) of the building. As has been pointed out in Chapter 2, it is the function rather than the form which has the greatest influence on cost because of 'social opportunity costing'. Hence construction economists find cost information identified in terms of function rather than form more useful, especially at the feasibility stage, as it provides us with more objectivity and reliability in terms of predicting costs. It is useful to note how the table relating to primary schools has a rather different set of values reflecting the different functions of the other two, namely, hospitals and factories.

Form is an important design variable but it is of secondary significance. In due course as more detail becomes known about the project, we will begin to investigate the influence of form on the likely costs and their distribution amongst the various elements. Later in this chapter, and addressed in great detail in Chapters 9, 10 and 11, we will discuss how we can take account of form such as shape, height, number of storeys and other design characteristics.

Figures 6.1–6.4 show a number of design, cost and statistical parameters that need further explanation so that the reader may interpret and use them on future projects. Considering Fig. 6.1:

■ $£/m^2$ *gross internal floor area*. This is the cost derived from priced bills of quantities making up the sample, in this case 10 divided by the gross external floor

Rate per m^2 gross internal floor area for the building excluding external works and contingencies and with preliminaries apportioned by cost. Last Updated 26-Jan-2002.

At 3Q2001 prices (based on a Tender Price Index of 186) and UK mean location.

Building Function	$£/m^2$ gross internal floor area						
	Mean	Lowest	Lower Quartile	Median	Upper Quartile	Highest	Sample
New build							
712. Primary schools							
Generally	869	372	717	854	996	1636	639
Up to 500 m^2 GFA	922	403	756	902	1067	1607	82
500 to 2000 m^2 GFA	854	372	698	837	979	1636	490
Over 2000 m^2 GFA	919	443	763	902	1054	1429	67

Figure 6.1   Average building prices: primary schools.

Functional unit rate for the building excluding external works and contingencies and with preliminaries apportioned by cost. Last Updated 26-Jan-2002.

At 3Q2001 prices (based on a Tender Price Index of 186) and UK mean location.

Building Function	£/functional unit						
	Mean	Lowest	Lower Quartile	Median	Upper Quartile	Highest	Sample
New build							
712. **Primary schools**							
m² usable floor area	1298	611	944	1229	1541	3077	110
No places	4262	1272	2813	4119	5121	18155	262

Figure 6.2 Average building prices: primary schools shown per functional area.

Rate per m² gross internal floor area for the building excluding external works and contingencies and with preliminaries apportioned by cost. Last Updated 26-Jan-2002.

At 3Q2001 prices (based on a Tender Price Index of 186) and UK mean location.

Building Function	£/m² gross internal floor area						
	Mean	Lowest Quartile	Lower	Median Quartile	Upper	Highest	Sample
New build							
412. **General hospital, GP hospitals, cottage hospitals**							
Generally	1191	462	954	1163	1340	2717	455
Up to 100 m²	1233	462	921	1167	1440	2717	138
1000 to 7000 m² GFA	1186	471	980	1188	1330	2161	241
7000 to 15000 m²	1137	650	957	1091	1330	2017	40
Over 15000 m² GFA	1128	703	984	1091	1200	1663	36
Public	1187	462	952	1163	1337	2717	433
Private	1271	500	1102	1259	1482	1924	18

Figure 6.3 Average building prices: hospitals.

Building Function	£/m² gross internal floor area						
	Mean	Lowest	Lower Quartile	Median	Upper Quartile	Highest	Sample size
275.5 Factories for electronics, computers, or the like	733	449	655	674	778	1148	10

Figure 6.4 Average building prices: factories for electronic computers or the like.

area (defined as the overall floor area within the building measured up to the inside face of the external walls) to give a cost per m^2 to the nearest £. Note the figures exclude external works, which have little bearing on the cost of the actual building, as such costs are related to the size and complexity of the site and not to the building placed upon it.

■ *The sample size.* This is the number of building projects that make up the sample that has been analysed.

■ *The mean.* This is the sum of the rates in the sample (£/m^2) divided by the total number making up the sample, in this case 10. This needs to be used with care, especially with a small sample, as in this case, as one or two extreme values can considerably influence the results. Note the table relating to primary schools has a much larger sample and therefore the influence of any extreme values will be of much less significance.

■ *The range (lowest and highest).* These are the extreme values, i.e. the highest and lowest found in the sample. As discussed above, they can create considerable distortion to the mean with the obvious consequences.

■ *The median.* This is the middle statistic and not the middle of the range. This figure, unlike the mean, is not as easily affected by any extreme values making up the sample.

■ *25% the lower quartile.* This is the rate at which 25% of the rates fall below and 75% fall above.

■ *75% the upper quartile.* This is the rate at which 25% of the rates fall above and 75% fall below. Clearly half of all the rates fall between the lower and upper quartiles. The inter quartile range is a more accurate reflection of the measure of the spread, as any extreme values are not included.

Further explanation of the use of this model to identify the cost bracket for our project can be found in Chapter 9, but suffice to say that from the minimal information of knowing the anticipated floor area of our proposed building, we can allocate a likely cost bracket for our project.

However, in using the above information there are three other adjustments to the model that are essential if our prediction of cost is to be realistic:

(1) time
(2) location
(3) specification.

Incidentally, these three adjustments must be made to all our cost data at whatever stage within the design process we are carrying out design cost management.

6.2.1   *Adjusting for time*

Adjusting for time is carried out by the use of an index which in itself is a statistical, deductive cost model which allows the construction economist to measure changes in cost and price, between various times. Note they measure the general change and not actual costs and prices and therefore have to be used with other cost and price information such as in the figures above. The general use of these indices is:

$$\% \text{ Change} = \frac{\left(\begin{array}{cc}\text{Value at date} & \text{Value at base date of}\\ \text{under review} & \text{source information}\end{array}\right)}{\text{Value at base date of source information}} \times 100$$

A typical application would be to bring to the present day the cost per m^2 of our school project shown on page 60:

$$\text{Cost/m}^2 + \frac{\left[\text{Cost/m}^2 \times \left(\begin{array}{cc}\text{Value at date} & \text{Value at base date of}\\ \text{under review} & \text{source information}\end{array}\right)\right]}{\text{Value at base date of source information}}$$

So, substituting in some figures, the cost per m^2 at today's price will be:

$$900 + \frac{[900 \times (175 - 169)]}{169} = 900 + 32 = £932/\text{m}^2$$

where
900 = cost/m^2 (£) at 1st quarter 2001
169 = tender price index at 1st quarter 2001
175 = tender price index at 4th quarter 2001.

Thus, to build such a project in the 4th quarter of 2001 we would expect our client to pay £932/m^2.

The other point to note is that the use of cost and price indices needs considerable care. Selection of the wrong index will lead to catastrophic results when measuring change in costs and prices. When we are preparing financial information on behalf of the client we must use a price index as this index measures changes in the prices the constructor is likely to charge us for the project. If we wanted to measure changes in the cost of the project to the constructor, reflecting the costs of the various resources and ignoring any allowance for profit (i.e. the market conditions) then we would use a cost index.

Figure 6.5 illustrates why it is so important to use the correct index where the different profiles of the cost and price indices can be clearly seen. As can be seen there are three indices shown, which reflect changes in prices and costs as indicated in the key. Note that the indices are all brought to a base of 100 as at 1985. The indices are occasionally re-based in this way, to enable the figures to be more easily used with current and future cost information. The graphs would have shown a very different set of values for the indices if they had been set at 100 in 1974!

As a general rule, in times of boom, one would expect to see an increase in the price indices whereas in a slump the opposite will occur. This is because, as one

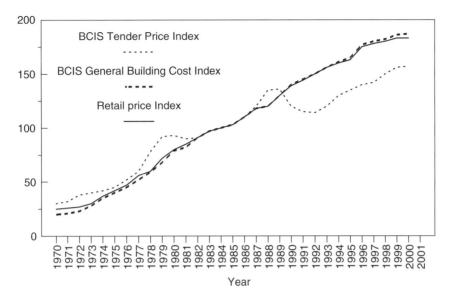

Figure 6.5 Cost and price indices.

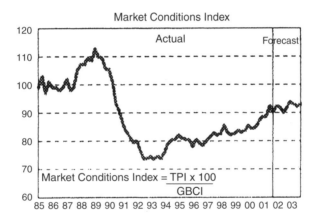

Figure 6.6 Market conditions index.

would expect, in times of a boom the constructor will be able to add more profit margin as there will be a surplus of contracts and not enough constructors around to carry them out. However, in times of a slump the opposite is clearly the case. From Figs 6.5 and 6.6 the reader can plot the times of slump and boom within the UK building industry. Notice in Fig. 6.5, the 1989 Tender Price index reflected the boom in construction, whereas the General Building Cost index did not mirror this trend. By 1992 this boom had turned into a slump, as reflected in the Tender Price index, whereas the cost of construction reflected in the General Building Cost index continued on a steady inflationary trend. It is also interesting to note that the construction industry does not always mirror the general trend in prices, as illustrated by the Retail Price index, which shows a general increase since 1990, following the trend of the General Building Cost index. Notice that there was an earlier boom in construction from 1978 through to 1980. Thus the cost indices tend to be influenced by inflationary or deflationary trends in the cost of resources. To highlight the market conditions pertaining to the construction industry the BCIS also produces a market conditions index as shown in Fig. 6.6. This graph indicates very clearly the booms and slumps that have occurred since 1985.

There are a number of further indices published by the BCIS as indicated below which reflect changes in both price and cost. For example, there are a whole series of all-in Tender Price indices which reflect changes in tender prices in different regions as shown in Fig. 6.9 and different client types, which are of assistance in attempting to predict the likely changes in price depending on the nature of the client and the region the project is to be built in. In addition, there are indices reflecting different costs of construction work and resources:

- General Building Cost
- Steel Framed
- Concrete Framed
- Brick
- General Building Cost (excluding mechanical and electrical)
- Mechanical and Electrical
- Labour
- Materials
- Plant.

The various cost indices are useful to construction economists in that they help to explain and allow us to predict, with more confidence, different trends in various types of construction so that advice can be given as to which form of construction would give the best value for money. Perusal of the BCIS indices for Brick construction and for Steel Framed construction, as shown in Fig. 6.8, indicate that from January 1995 until October 1998, the steel framed buildings cost trend is slightly more than that of brick construction. However, since 1997 the opposite is the case. That is, it is more expensive for the constructor to build in brickwork than steel. This can be perhaps explained by studying Fig. 6.6 for the BCIS index of all-in hourly rates since 1998, which indicates a considerable increase in labour costs. However, there has been no increase in material costs over the same period, as indicated by the DETR construction materials cost index (Fig. 6.7). As brickwork is more labour intensive than steelwork then, as shown by the indices, all other things being equal, we should advise our client to build in steel. Incidently, the reason steelwork is so widely used in the USA is because site labour in the USA is much better paid than the equivalent factory worker. The productivity of the site steel fabricators is high and as steel is essentially a factory-based prefabricated process, it is a cheaper option than masonry work. In the UK construction workers are paid less than their counterpart factory workers, which partly explains why prefabricated construction generally is used much less in the UK than in the USA.

Input Cost Indices

(Base 1985 = 100)

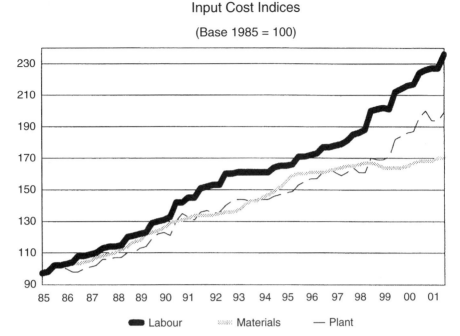

Figure 6.7 Input cost indices.

To allow us to make adjustments for changes in price we have included the BCIS all-in Tender Price index, covering the necessary time periods. We have also included, on the website, the General Building Cost index covering the same time periods so that the reader can see the difference in the cost information the application of these two indices would make to any chosen project. When using these indices it is important to remember that they are derived from historic information. The all-in tender-based price indices are prepared from sample bill rates in bills of quantities which are statistically analysed to ensure their reliability. These indices are generally based on the Property Services Agency (PSA) *Schedule of Rates for Building Works* (Property Services Agency 1990). This schedule of priced bills of quantities rates is very comprehensive and provides adequate coverage for the majority of buildings.

The factor cost indices, on the other hand, are constructed by analysing a typical building into constituent proportions of the basic resources: labour, mater-ials and plant. The various BCIS building cost indices listed above are based on the *Price Adjustment Formulae for Construction Categories (Series 2)* (Department of Employment, Trade and Resources 1993). The Price Adjustment Formulae are used to adjust for increases or decreases in costs, usually at the monthly valuation stage, in contracts where fluctuations are allowed.

The civil engineering sector has a similar formulae driven approach the description of which can be found in *The ICE Conditions of Contract* (Institution of Civil Engineers 1999).

Base: 1985 mean = 100

Quarter		BCIS Steel Framed Construction Cost	BCIS Brick Construction Cost
1992	i	143	143
	ii	144	144
	iii	147	147
	iv	148	147
1993	i	149	148
	ii	151	149
	iii	152	151
	iv	153	151
1994	i	154	152
	ii	156	153
	iii	158	156
	iv	159	157
1995	i	162	159
	ii	164	162
	iii	167	165
	iv	168	165
1996	i	169	165
	ii	169	166
	iii	171	168
	iv	171	168
1997	i	172	169
	ii	173	171
	iii	174	172
	iv	176	174
1998	i	177	174
	ii	178	176
	iii	182	181
	iv	182	182
1999	i	181	181
	ii	181	181
	iii	184	186
	iv	185	187
2000	i	186	188
	ii	188	190
	iii	192	194
	iv	193	195
2001	i	193	195
	ii	194	196
	iii*	197	200
Forecast	iv		
2002	i		
	ii		
	iii		
	iv		
2003	i		
	ii		
	iii		
	iv		

*Provisional.

Figure 6.8 Steel Framed construction and Brick construction cost index.

These procedures were developed to take the place of the more laborious and time-consuming approach, based on keeping records of labour, plant and material resource costs during the progress of the work. The constructors were then asked to declare in their tender submission the costs of these basic resources, used in the preparation of the tender. From this information the increases or decreases in costs of these basic resources could be laboriously calculated.

As can be seen from the brief description above, their construction is complex. For a detailed explanation of how they are prepared the reader is recommended to consult the literature (Morton and Jaggar 1995, Ferry *et al.* 1999).

It is also important to be aware that the various indices always lag behind cost and market trends. As a result the latest figures in the index are recorded as provisional rather than firm, indicating that they may be subject to further review and change. Additionally, the construction project we are interested in building will always be in the future, and therefore, we have to resort to forecasting rather than using firm figures from the index. Perusal of the indices shown in Fig. 6.8 and on the website, highlight which index values are firm or actual, which are provisional (*) and which are forecast.

The use of any of these figures from these indices needs caution, especially for use in forecasting. As pointed out above, when discussing deductive models possible trends may prove entirely incorrect as they may be influenced by entirely unforecastable events, such as the attack on the World Trade Centre twin towers in New York on 11 September 2001 which led to a number of unanticipated economic events, as for example the downturns in tourism and the airline industry.

In addition to the BCIS indices described there are a number of other published indices, as highlighted below, which are made use of by the building industry:

- The Building housing cost index which is produced by the *Building* magazine and, as the name suggests, relates to housing.
- The DETR (Construction sponsorship Directorate) index of building tender prices. This was the original Tender Price index from which most of the other tender price indices have been developed.
- Davis, Langdon and Everest Tender Price index. This index is published quarterly in the *Building* magazine.

To summarise, these various indices are of benefit to the construction economist in the following way:

- Updating of project prices in design cost management
- The extrapolation of existing trends into the future
- The identification of changing cost relationships such as might occur in different types of construction
- The calculation of fluctuations in contracts at monthly valuations and at the final account stage
- The assessment of market conditions and their implications on the cost the client will have to pay for his or her building.

6.2.2   *Adjusting for location*

The second area of cost influence we need to take into account when establishing the cost bracket at the feasibility stage is that of location. Again, this adjustment must be taken into account at every stage in the development of the design of the construction project. Clearly, location both at the regional (macro) and the site (micro) level will have considerable influence on the likely financial implications of construction projects. At international level, the cost of construction will vary enormously, depending on a variety of factors such as the economic, political and legal conditions of the particular country such as availability of resources, levels of skills, availability of technology, degree of risk, infrastructure, financing and communications.

Returning to the UK, a series of location-based models have been developed by the BCIS, which allow relatively fine adjustments to be made to our cost data. For further information, see the literature (The Royal Institution of Chartered Surveyors 1992). Figure 6.9 shows the regional indices.

The website accompanying this book contains a sample of such indices which can be used to take account of the different locations of the source data, as well as the construction project under consideration.

The BCIS location indices work by assuming a national average of 1.0. In fact the $£/m^2$ shown in the tables for hospitals, factories and primary schools are based on the national average of 1.0 which may then need to be adjusted to take account of the particular location of interest. This is carried out as shown below:

$$\% \text{ Change} = \frac{\left(\begin{array}{c}\text{Value at location} \\ \text{under review}\end{array} - \begin{array}{c}\text{Value at location of} \\ \text{source information}\end{array}\right)}{\text{Value at location of source information}} \times 100$$

$$\text{Cost/m}^2 + \frac{\left[\text{Cost/m}^2 \times \left(\begin{array}{c}\text{Value at location} \\ \text{under review}\end{array} - \begin{array}{c}\text{Value at location of} \\ \text{source information}\end{array}\right)\right]}{\text{Value at location of source information}}$$

So, substituting in some figures, the cost per m^2 today, taking into account our new location will be:

$$932 + \frac{[932 \times (1.08 - 1.0)]}{1.0} = 932 + 74 = £1006.5/m^2$$

where
932 = cost/m^2 (£) at 4th quarter 2001 (as adjusted above)
1.08 = location index of our proposed project
1.0 = location index of source information (as adjusted by the BCIS to national average).

Figure 6.9   UK regional indices.

Thus, to build such a project in the 4th quarter of 2001, in our proposed location, we would expect our client to pay within a range around £1006.5/m².

Of course, if our location index is less than the source information index, then the cost of the project will be lower, all other things being equal.

Thus, these indices allow us to take account of regional variations within the UK and can be used to make broad adjustments for differences in location. Again, the indices are compiled from information derived from sample priced bills of quantities, which after statistical processing are made as reliable as possible. However, great care must be taken when they are being used. For example, we know that it is much more expensive to build in central London than the north-east of England because of more restricted site working as well as more expensive labour rates prevail in the London area (Fig. 6.9).

When we begin to focus on the actual site itself, there may be all kinds of individual anomalies, which will considerably influence the costs such as:

- topography
- proximity to other buildings and structures
- existing site services
- green- or brownfield site
- contaminated land
- particular planning restrictions, i.e. height, etc.
- proximity to water, gas electricity supplies, etc.
- restrictions, such as parking, noise levels, etc.

Clearly all these factors will have considerable influence on the cost of the project, hence the construction economist will need to use considerable judgement when carrying out design cost management, especially at the feasibility stage of the RIBA plan of work, when very little information will be known about the project.

To illustrate the problem of adjusting for location, an example from the civil engineering industry will serve to make the point well. The construction of a hydroelectric power station, in the middle of the highlands of Scotland, at first sight would appear to be in a low cost location, as indicated by reference to the regional indices shown in Fig. 6.9, because of lower labour costs and the unrestricted nature of the site. However, such a conclusion would be erroneous, as part of the site will undoubtedly be at the top of a mountain, entailing very high installation and removal costs, such as very expensive service roads, power and utilities up the mountain side. Therefore, the use of the BCIS location index would give entirely the wrong result if it was not considered with local information. In fact, in a local government authority, where one of the authors worked in the late 1960s, there was a very astute Chief Quantity Surveyor, who was able to claim considerable extra funding from central government for the authority's school programmes. Often, because of the National Park planning restrictions, many of the schools had to be sited some considerable distance from the near by highway, thus resulting in extra infrastructure costs due to longer access roads, increased lengths of service runs, landscaping and other site specific costs.

6.2.3   *Adjusting for specification*

The next adjustment we need to make to our cost model is that of specification, which, as we move through the design process, will become more and more comprehensive, as we interpret and develop the initial performance specification, set by the client and interpreted by the design team. However, at the feasibility stage, all we are likely to know is the functional performance of the project and some concept as to the likely size. We may also have some information concerning outline planning restrictions such as height limitations, plot ratios, required external appearance (for example, natural stone walling in a national park or Georgian style windows in a terrace of existing Georgian properties).

Additionally, it is necessary to make some allowance for the external works as the costs/m^2 relating to different building types, as shown in the figures above, for example, do not include an allowance for external works. That is, all work beyond the outside face of the external wall of the building(s). This may seem slightly surprising at first but, as suggested above, the cost of the external works does not have a direct bearing on the cost of the building. The cost of the external work will depend primarily on the size, shape and topography of the site, together with the distances to, and availability of, existing services. Clearly some of the external works' requirement will be specific to the functional needs of the building, such as playing areas for schools, where mandatory requirements are laid down. The planning restrictions may also provide certain requirements such as a specific plot ratio and perhaps the need to provide, for example, a certain minimum level of car parking. Thus, the construction economist will have to make some kind of allowance for external works based on experience and judgement at this early stage.

Thus, the work of the design team including the construction economist is to interpret and develop the client's performance specification, in order to ultimately convert it into a definitive specification: the finished project.

Chapter 9, together with the accompanying website, demonstrates and shows the process of establishing a cost bracket or range based on minimal information as discussed above.

6.3   Summary

A cost bracket is established by collecting broad cost information (£/m^2) of buildings of the same function, as for example available on the BCIS, or from cost information related to the performance criteria required such as £ per cost place for pupils in a school. This information is then adjusted for size, time using a tender price index, location using a location index and finally for specification, as well as taking account of any known specific information about the project such as the nature of the site, any restrictions or any specific requirements. At this early stage it is essential to set the financial boundaries, whilst not making a premature decision on a specific solution in order to continue to seek the single best solution to the client's requirements as shown in Fig. 1.4. It is also possible that the financial boundary, through the application of the Developers Equation, will be set by the client and a proposed solution will emerge within this financial constraint in terms of what physical space and specification can be provided.

As can be seen, through the application of the BCIS database, we have used primarily deductive cost modelling techniques, because by using the BCIS statistical analysis, the cost information has been made as reliable and robust as possible. However, in such applications we must use care and great judgement. Thus, considerable experience is required from the construction economist to arrive at solutions that set the framework within which to develop an optimal solution to the client's requirements.

6.4 Reader reflections

- How may statistical deductive models be used to develop a cost bracket at feasibility stage?
- How do these models represent uncertainty within the data?
- Why must the construction economist select the correct index? What is the difference between the indices available?
- How has the market for construction influenced tender prices over the last 20 years?
- When would we use local knowledge to override the locational index information given by the BCIS?

7 Design cost management: sketch plan stage

7.1 Introduction

This chapter aims to:

- Introduce the reader to the processes used by the construction economist at the sketch plan stage
- Identify the type of cost information stored in outline and detailed elemental format
- Define the design features that may be taken into account at this stage
- Demonstrate adjustments to be made
- Discuss the concepts of price and design risk.

Having established the likely cost bracket or range at the feasibility stage the design team can begin to develop the design in more detail as we move into the sketch plan stage. As shown in the RIBA plan of work in Fig. 5.3 there are two subsets of the sketch plan stage:

(1) outline proposals
(2) detailed design.

The sketch plan stage is where the construction economist begins the process of producing the first structured cost plan, which aims to confirm the budget set at the feasibility stage. Equally important, this first cost plan allocates, to each functional element making up the project, the appropriate amount of money, in order to achieve the optimal solution for that element, within the overall context of the project. In order to carry out this process more detailed inductive models of the building will be used, based on, wherever possible, algebraic relationships to establish the various cost allocations. It is likely that a number of iterations will be carried out, in order to gradually develop the project cost model, which will initially exist as an outline cost plan at the outline proposals stage before it becomes a more comprehensive cost plan at the detailed design stage. The detailed cost plan is confirmation of the budget and an accompanying statement of how much money is to be spent on each detailed functional element.

Chapters 9 and 10 and the accompanying website explain, in detail, the establishment of the outline and detailed cost plans, whilst also allowing the website user to perform his or her manipulations in order to establish a typical outline proposals cost plan.

It is, however, worthwhile to briefly discuss the processes that the construction economist needs to undertake in establishing these cost plans. During this stage, the design team will be beginning to firm up their proposals in terms of shape, number of floors, levels of specifications and other important design information. In order to build the cost model that reflects these decisions, they will usually access, initially, outline cost analyses which match as closely as possible the project under consideration. This 'closeness' should be matched in terms of function and thereafter, in as much detail as possible, to the project under consideration. This will include for example, size, level of specification and location. Additionally, the closer the analysis or analyses are to the project in terms of time, so much the better.

This is helpful for two reasons:

(1) the nearer in time the more likely similar economic conditions will prevail
(2) to ensure that the technological, legal, planning, changes in recommendations and procedures, etc. relating to the analysis or analyses are as close a match to the project under consideration.

Clearly, the further in time the analysis or analyses are from the project, then the less close the match will be and thus the less reliable the ultimately built cost model will be. Finding a perfect match is virtually impossible and, as suggested above, invariably it is often difficult to find such matches when the building is unique. For example, there are very few cost analyses of opera houses available. The construction economist will naturally rely on their own analyses initially, if any are available from their own sources, as they will place more confidence on information they are familiar with. Unfortunately the chance of having such information 'in house' is unlikely. As a result they are likely to search the BCIS database on a variety of search criteria, as discussed in Chapter 10 and demonstrated on the website in order to 'home in' on the closest match possible. Once the selection has been made, the analyses, if more than one has been found, can be adjusted for size, time, location and specification as discussed above.

7.2 Elemental cost information

Examples are shown in Figs 7.1 and 7.2 of typical group and elemental cost analyses which are good matches for our project which, of course, are cost models of the historic projects they represent. Currently some of the BCIS elemental analyses are accompanied by outline plans and elevations which is extremely helpful in providing the construction economist with a rapid appraisal of the nature of the project described in the cost analysis.

This information is particularly useful, as it gives the construction economist a rapid insight into the form, arrangement and appearance of the building, as well as allowing some physical dimensions of the building to be obtained.

These analyses can now be adjusted for size, time, location, and specification, as detailed in Chapter 10. To study these analyses in more detail visit the website.

Log on to the website to review cost analysis.

Production facility for electronics

Date: 29 September 1999   Analysis: 12324

Floor area: 5450 m^2, 1 (2) storey

Electronics production facility. RC ground slab, upper floor and stairs. Steel frame. Facing block and curtain walling; metal cladding to walls and steel flat roof. Double-glazed aluminium windows; revolving doors; roller shutters. Block, metal stud, timber stud, relocatable and cubicle partitions. Flush doors. Plaster, plasterboard and paid to walls; vinyl and carpet flooring; suspended ceilings. Fittings, sanitary ware. Gas LTHW central heating, air conditioning, ventilation, electrics. Lift. Kitchen equipment; rainwater recycling; lightning protection; alarms.

Total cost: £2 892 341   Cost/m^2: £530.70   Group elemental analysis

Figure 7.1   Group elemental analysis summary.

Factory, Site 4

Date: 30 July 1997   Analysis: 23456

Floor area: 1850 m^2, 1 storey

Two factory units for computer-related industries comprising a two-storey professional block and single-storey assembly block with integral two-storey offices. RC ground beams, piling, ground slabs. Steel frame with insulated metal roof and wall cladding. Double-glazed aluminium windows. Steel stairs. PCC upper floors. Block and metal stud partitions. Timber internal doors. Plaster paper and paid to walls; vinyl, tiles, rubber, carpet and raised access floors; mineral fibre suspended ceilings. Fittings. Oil fired central heating, air conditioning, electrics and lift.

Total cost: £1 319 301   Cost/m^2: £713.14   Full elemental analysis

Figure 7.2   Detailed elemental analysis.

These adjusted analyses are, in essence, crude cost models of the proposed project under consideration. Further adjustments for size and specification are not quite as straightforward, as the process necessitates going through the various elements in turn and taking account of anticipated differences in size, including geometry, and specification, and adjusting as necessary. This process will often be carried out in stages, the first stage being based on the group elemental format, during the outline proposals stage and detailed elements, during the scheme design stage of the RIBA plan of work.

Full details of this process are described in Chapter 10, including the various adjustments and are demonstrated on the website. The reader is reminded to visit the website to study the BCIS standard form of cost analysis (SFCA) which gives the guidelines, principles and definitions in the production and use of elemental cost analyses. This is also shown in Appendix 2.

Go to the website to view the standard form of cost analysis.

The reader should study, in particular, the definitions of each functional element to ensure they fully understand what is and what is not included in each element, as well as explaining and defining the numeric information, such as *gross floor area*, *element unit quantities* and *element unit rates*.

Summarised below are the various modelling applications that can be used to help identify the shape, size and specification of the project in order to establish a realistic cost model:

- *Gross internal floor and cost per m²* have already been discussed in Chapter 6, but their strict definitions can be found in the SFCA. Additionally, perusal of the analyses show that each element is also accompanied by the cost per m². This allows us to allocate the costs of the various elements, with more accuracy, as explained in Chapter 10.
- *Element unit quantities and element unit rates*, the definitions of which can be found in the SFCA. Observation of the cost analyses shows that against some of the elements, element unit quantities and element unit rates together with specification details have been included. This information allows us to make a much more accurate prediction of the cost target for each element, as the element unit quantity is a measure of how much of that particular element there is in the analysis. With the accompanying element unit rates and specification information, as demonstrated in Chapter 10 and on the website, we can arrive at a detailed cost target for the element under consideration. A particular problem is, unfortunately, that the element unit rates and quantities are often not detailed on the cost analyses as the examples here indicate, thus impeding the building of our detailed cost model describing the project.
- *Number of storeys*. Generally multi-storey construction is more expensive than the equivalent floor area on one floor. A number of regression models have been built which give information on height (the independent variable) and cost per m² (the dependent variable). This model can be viewed on the website.

Go to the website to view regression analysis on height and cost.

- *Wall/floor ratio*. Again this is defined in the SFCA. The wall/floor ratio is a measure of the efficiency of the enclosure of space. The smaller the enclosing wall area compared with the gross floor area the more efficient the shape of the building.
- *POP ratio*. This again is a measure of the most efficient shape of enclosure (actually a circle) but, because of the complexity of producing curved materials and components, together with more complex setting out of problems, a square plan shape is the most efficient shape in terms of enclosure.

The POP ratio is expressed as a percentage (%). The higher the percentage the more efficient the shape of enclosure. The formula is:

$$\% \text{ Compactness} = \frac{2\sqrt{\pi A}}{P} \times 100$$

where A is the fully enclosed covered area of a typical floor level and P is the perimeter enclosing that area, measured internally.

- *Storey enclosure method.* Of historic interest only, but was an early cost model which attempted to take account of shape and height in a simple formula.
- *The cube method.* Again, only of historic interest. It relied on keeping records of the cost per cubic foot of different buildings. The volume of the proposed building in cubic feet could then be calculated and a cost target established from the cost per cubic foot for that particular building type.

Details of these various modelling aids can be found in the literature (Flanagan and Tate 1997, Ferry *et al.* 1999, Smith and Love 2000).

Thus, the construction economist can, with the aid of these models including, where available, graphical information, construct a working cost model of the construction project which, at the end of the scheme design stage, will have established cost targets against each of the detailed elements and will have confirmed the budget established at the feasibility stage. Remember this detailed cost model, in the form of a detailed cost plan, is not just about confirming the budget, but is also ensuring that the design team have established the optimal solution for the project, in terms of the clients performance specification, expressed through the functional elements making up the project. The website shows the developed detailed cost plan for our particular project.

Log on to the website to browse the cost plans for the case study.

7.3   Completing the adjustments

As described above, much of the financial allocation is based on varying levels of mathematical manipulation, to take account of time, location shape, size, height, etc. and they therefore tend to involve objective decision making. However, much subjective decision making is also involved to take account of the specification needs and relies heavily on the experience and judgement of the construction economist. This experience and judgement is particularly important in the unique element of Preliminaries. (Review the elemental definition in the SFCA.) This element is unique because it forms no lasting part of the hardware of the project, but is a financial device to allow the construction economist (and the contractor) to establish a cost target reflecting the anticipated financial implications of the construction company managing the project, together with the recovery of the indirect costs associated with the running of the construction business. The Preliminaries form an integral part of a bill of quantities and allow the constructor, as part of his or her estimating process to price for these costs. Thus, the element of Preliminaries forms a natural element within which to analyse and predict such costs. The problem is, as highlighted above, that the constructor can virtually price this part of the bill in any way he/she might choose. For example, the Preliminaries may be allocated to and priced within the measured quantities contained in the rest of the bill, or alternatively be priced entirely realistically within the Preliminaries section of the bill of quantities. This was highlighted earlier and the rules of the Standard Method of Measurement have now been drafted in a way that encourages this more realistic approach to their pricing. It is not the purpose of this book to explain and describe the various approaches to the derivation of and the pricing of the Preliminaries, as there are many books which detail this process (Brook 1993, Smith 1986).

In practice the constructor will generally use pricing strategies, ranging between these two approaches, with the objective of improving cash flow manipulations as noted in Chapter 5. The main point to make is that the allocation of a cost target for the Preliminaries is rather subjective and therefore difficult to predict with any reliability. The only thing one can say is that, within the overall price for the project, as reflected within the cost analysis, somewhere these indirect costs will have been included, together with the inclusion of profit, as the tender figure reflects the market price for that particular project. For this reason the detailed cost analyses are shown as two presentations: one with the Preliminaries shown as a separate element; and the other with the Preliminaries distributed on a per-centage basis throughout the remaining elements.

This information is helpful where the construction economist is using two or more analyses to produce the cost model of the project. What are known as the vagaries of pricing the Preliminaries can be partly mitigated against by deducting from the all-inclusive elemental costs the likely allowances for the Preliminaries, in an attempt to establish the correct financial benchmark for the analyses under consideration. The construction economist would tend to use more than one analysis to help develop the cost model of the Preliminaries where one analysis may not embrace all the elements under consideration, especially in specification terms. The economist might also look at more than one analysis to gain more confidence in the establishment of the various elemental targets. For example, the cost analyses may contain incorrect elemental distributions, either through error or deliberate pricing distortions by the constructor. One can now see why using analyses prepared by the construction economist's organisation is so much more reliable, as the quality and content of the information is known. Thus, by using this strategy, a realistic allocation of a target for the Preliminaries can be estab-lished for the project under consideration. Chapter 10 and the website demon-strate this process in more detail.

It is generally recognised, and expected, that deterministic techniques will be used in the prediction of the costs of construction, but in truth it is an entirely unrealistic expectation. The notion of a single figure estimate for a construction project during the design stage is utopian indeed and, as a result, construction economists have developed techniques to compensate for this fictitious expect-ation. It is not unusual to find students of construction economics proudly pro-ducing budget estimates with a final figure expressed to the nearest penny never mind the nearest pound!

Thus, in the establishment of the outline and detailed cost plan the construc-tion economist will include an allowance for Price and Design Risk. In fact, these are allowances in the form of a contingency to assist the construction economist manage the financial risks associated with deterministic budget estimating.

The Price Risk is a percentage allowance to take account of likely increases in the budget between the date at which the budget is established and the date at which the project is likely to go out to tender. The percentage allowance will be based on the anticipated changes in market conditions that might occur during the time interval between the two dates.

Design Risk is an even more uncertain percentage-based allowance, which is included to take account of uncertainty in the design process in terms of possible changes or departures from the anticipated solution. This allowance reflects the familiarity of the design team with the nature of the project under consideration as well as the confidence the design team has in each other. For example, a team that has worked successfully and closely on a similar project to that now being considered, will have a low Design Risk, whereas the opposite will prevail where such certainty is not present. When the establishment of the final cost model has been achieved, one would expect to see that the Design Risk has disappeared completely as all the uncertainty associated with the design solution will have been resolved. The Price Risk will still remain, although this will be further refined, due to the closing of the two dates at which the final detailed cost plan has been established during the Working Drawing Stage of the RIBA plan of work (RIBA 2000).

The inclusion of Price and Design Risk, as described above, reflects the recognised approach to the process of design cost management in terms of establishing the budget, producing the cost plan and carrying out the cost checking process described below. It is also usual to include a separate Price Risk, as a percentage, to take account of the time between the date of going to tender and the completion date for the project. Of course, this allowance will depend on the anticipated duration of the construction phase and whether or not fluctuations have been included under the particular contract conditions that are applicable. Obviously, the longer the contract period, the greater the chance of a change occurring in the likely future price. Where no fluctuations are allowed the constructor bears this risk and, where they are allowed, the client carries the risk. The transfer of this risk should be carefully considered as in times of a boom, with an excess of work, the constructor will be able to add considerably to allow for fluctuations, which may ultimately cost the client far more than if fluctuations had been allowed for in the contract. Of course the reverse of this scenario is true in times of a slump. The reader is reminded of the benefit of the price and cost indices discussed earlier in this chapter to the construction economist in helping to establish the necessary allowances for Price Risk.

7.4   Reader reflections

- Review the specification details of the two projects shown in this chapter.
  (1) What are the similarities and differences?
  (2) How might this influence the costs of these projects?
- What do you understand by Price Risk and how would the construction economist allow for it?
- What is understood by Design Risk and how would the associated costs be allowed within the budget at this stage?
- How does the approach to the consideration of the Preliminaries element vary from other elements in the cost plan?

8 Design cost management: working drawing stage

8.1 Introduction

This chapter aims to:

- Introduce the reader to the processes used by the construction economist at the working drawing stage
- Identify the type of design information produced by the design team at this stage
- Introduce the reader to the difficulty in establishing the validity of the cost targets for specialist works
- Introduce value engineering terms.

The final stage of the RIBA plan of work is the working drawing stage which comprises three subsets:

(1) detail design
(2) production information
(3) production of the bill of quantities.

This stage is of great importance in that, during the production of the working drawings, the targets established at the sketch plan stage are checked and re-checked as design decisions are finally made. The checking process will tell us three important possible outcomes:

(1) Confirm the financial target established for a particular element, hence making no change necessary to the design of that element.
(2) Confirm that the financial target for a particular element has not been met. This means that the size, arrangement and or specification of that particular element can be increased without impinging on the budget.
(3) Confirm that the financial target for a particular element has been exceeded. If this is the case then the cost of the element will have to be reduced in some way, by considering its size, arrangement and specification or any combination thereof and making some adjustment. Alternatively, any savings from any other elements can be allocated to this particular element thus ensuring that the overall budget is not compromised.

Although the above scenarios seem simple and straightforward, the expertise of the construction economist will play an important role in reviewing all the elements, in order to fully consider their various implications so as to be able to decide what adjustments, if necessary, need to be made.

The cost checks are made by deploying the most accurate and detailed form of cost model available, apart from the constructor's unit rates, namely *approximate quantities*. This approach is based on amalgams of units of finished work that bear the same unit of measurement, such as the substructure where a combination of items, for example, of excavation, hardcore, concrete slab and reinforcement can be given an all-in unit rate and the area can be calculated from the drawings.

Go to the website to review the format and content of the various elemental cost checks.

The website and Chapter 11 show this approach in detail. The reason that this strategy can be deployed is because as the working drawings are produced, together with accompanying specification information, there is sufficient detail to be able to establish a more accurate and reliable cost target for each element. Thus, by this process we are able to perform cost checks on the various targets that have been established in the detailed cost plan, produced during the sketch plan stage of the RIBA plan of work. This enables the construction economist to control the financial status of the project—an essential prerequisite of any financial management system.

Once the cost target has been validated in this way, the design team can now proceed to generate production information and bills of quantities as a basis for tendering and the selection of a suitable constructor to carry out the construction work.

When the tender bids are returned the construction economist will receive confirmation of the degree of success of the design cost management that has been deployed for the particular project under consideration. If the design cost management has not been carried out correctly, then time-consuming remedial action will be necessary which often results in a less than satisfactory solution, as the potential to make changes at this stage tends to be limited to that of a more cosmetic nature, such as the substitution of different internal finishes. It is also worth pointing out that tenders coming in under budget can also be regarded as unsuccessful as a suboptimal solution will have been achieved. Neither the design team nor the client is likely to be pleased at this outcome.

The priced bill of quantities submitted by the constructor will complete the circle, and in due course will become a cost analysis for use in future cost modelling exercises.

8.2   Review of the process

We have now traced the nature of the cost models that are created and used, from feasibility through to the working drawing stage of the RIBA plan of work. These models are based on the necessary design cost management tasks of:

- establishing the budget
- preparing a detailed cost plan

- confirming the robustness of the cost plan
- taking any necessary remedial action should discrepancies between the budget and the cost plan occur.

We have shown that market driven prices at very coarse levels, such as costs per m^2, through to detailed elemental cost analyses, must all be adjusted for size, time, location and specification to form the basis of the various cost models tracing the financial implications of the project under consideration.

It is important to remember that these are iterative processes aimed at directing the design solution towards the optimal construction solution to the client's performance requirements (needs). A further point worth emphasising is that some of the elements are more significant than others, both in financial terms and because their ultimate solution may impact on other elements. For this reason such elements will need to be considered as a first priority and will require more detailed attention than some relatively minor elements. The construction economist will be well aware of which elements are of significance and how they might impact on other elements and again these issues are addressed later in the book. However, an obvious example is the element of external walls, the design of which will have a major impact upon the remaining structural elements. Conversely, an element such as ceiling finishes will have little influence on other elements, itself being influenced by the service elements, as well as structural elements, such as upper floors.

A particular constraint in the establishment of our cost models is a significant limitation in the availability of financial data concerning the key elements of services. These elements are complex and may account for up to 40% of the total cost of a highly serviced project such as a hospital. Additionally, they are of critical significance in terms of how they are incorporated in the project and as a result have a major impact on the constructor's programming of the project, and thus have a major impact on its buildability. Unfortunately the cost of services is generally included in the form of an all-embracing prime cost (PC) sum, which is an educated best guess by the consultant service engineers as to their likely costs. Therefore, no detail is provided of the individual costs of the constituent parts of these various service systems. Although the rules of the Standard Method of Measurement allow for the description and quantification of such work they are seldom applied. The latest edition of the Standard Method also recognises the difficulty the constructor faces in trying to take account of their installation, within his or her programme, by the introduction of *defined* and *undefined prime cost sums* (CPI 1998).

A defined prime cost sum must be accompanied by sufficient additional information, such that the programming implications can be taken into account during the tendering stage. An undefined prime cost sum means that the design team is not in a position to be able to provide sufficient information to allow the constructor to incorporate it into the programme. It introduces a greater element of risk into the project for the constructor to price and allow an adequate amount to deal with the situation. This situation is also not very helpful to the client as it means that the constructor can seek additional payment for the programming implications, once their detail is known, at some point during the carrying out of the work on site. Despite these improvements the detailed measurement of services remains infrequent for the following reasons:

- Traditionally, the use of prime cost sums has always been applied and is a straightforward and simple mechanism to include services in the bill of quantities.
- Services are complex, and in order to measure them a detailed knowledge of such systems is required, which many quantity surveyors do not possess. It can also be argued that their measurement as a basis for cost allocation is fairly futile in that the costs are performance related, rather than quantity related. To illustrate the point a boiler with an output of 40 kW is unlikely to cost twice as much as a 20 kW boiler. However, a doubling of low cost supply pipework, the cost of which is quantity related, will cost about twice as much.
- High level services are probably better specified by performance, rather than as a definitive specification. This is because many alternative solutions are available and therefore, in order to seek the best solution, maximum innovation can be achieved by the use of performance related criteria.

Thus, the cost modelling of the service elements remains a major problem due to the lack of reliable data for the reasons discussed above.

From this brief treatise of design cost management it can be seen that it is a complex and difficult process and, due to the high expectations of accuracy, it purports to achieve levels of accuracy that are not entirely realistic. As already commented upon, the source data are derived from priced bills of quantities and lack objectivity due to their opacity and the opportunity for the manipulation of the priced information by tendering constructors. Nevertheless with experience and judgement, design cost management as described above (amplified in Chapters 9, 10 and 11 and demonstrated on the website) has and does achieve excellent results in helping to achieve well balanced project solutions on behalf of the various clients.

In recent years the process of design cost management has been aided by the introduction of *value management*, sometimes termed value engineering, especially in the USA, where the technique was originally developed, although the take up of this useful aid to the design team remains limited. There are a number of definitions of this technique, with perhaps the most meaningful and useful being:

> The systematic application of recognised techniques which identify the function of a product or service, establish a monetary value for that function and provide the necessary function reliability at the lowest price.
>
> (Mudge 1971)

Clearly this definition embraces much of what cost modelling is about as developed and used by the construction economist. However, there is an important philosophical difference in that value management embraces the fact that a certain amount of unnecessary cost is inevitable in construction design due to the complexity of the process. So, value management concentrates on achieving value rather than just focusing on cost, value being assessed by considering the cost of components in relation to their function. In other words, cost is determined by what a component is, whereas value is determined by what a component does. Value management should not be seen as a cost saving exercise, but as a proactive process concerned with removing unnecessary cost, and more importantly, improving quality. Design cost management and value management can be integrated for the client's and design team's benefit. However, it is not the purpose of this book to explain how value management is carried out as there are many texts which deal extensively with this important technique (Mudge 1971, Kelly and Male 1993, McGeorge and Palmer 1997). The value management approach relies on taking a 'second look' at the key decisions, in a formal and structured way, in order to generate alternative proposals and test their validity. Typical examples might be the size and arrangement of supporting columns, which might be redesigned to create more uninterrupted space. The Thames barrage was a particular project that benefited from the use of value management. Value management therefore is a useful tool for all those concerned with the design and construction process, and is of particular benefit to the construction economist.

8.3 Summary

To date no one has come up with a better approach to the carrying out of design cost management as described in these chapters, despite its many limitations. In Chapter 12 we will suggest how things might change in the future, especially because of, as already discussed in Chapters 3 and 4, changes in procurement strategies which are gathering increasing momentum.

8.4 Reader reflections

■ How might the prime cost sum be arrived at if the design team were considering nominating a specialist contractor?
■ What skills would a value engineer possess in order to make a contribution to the design development?
■ How would design information become available in order to carry out the cost checks?
■ How might the design team respond to advice that the project was over budget? How would the construction economist formulate such advice?

9 Estimating the cost bracket: feasibility stage

9.1 Introduction

This chapter aims to:

- Introduce the reader to typical design information that is available at the briefing and feasibility stages
- Present the case study that is to be used to demonstrate how the principles discussed in the earlier chapters are applied
- Illustrate how a developer's equation can be used to arrive at a budget for construction
- Demonstrate the use of average building cost data when arriving at a cost bracket
- Highlight the areas of uncertainty in design and cost information that need to be taken into account at the early stages of a project.

9.2 The case study

9.2.1 Design information

An outline client's brief and detailed client's brief are also provided and should be consulted at the relevant stage in the cost planning process to replicate as far as possible the likely information the construction economist would have available. The drawings are provided as Acrobat portable document format files. They may be printed off in hard copy, however, care should be taken not to scale directly from the printed image as the scales shown are those of the original and are not accurate. The authors have attempted to replicate the increasing amount of design information that would be available as the project requirements become more specific. A site investigation report and borehole information are also included to give the reader an impression of the diversity of information that can inform the construction economist's decision making.

9.2.2 Cost information

The BCIS have given the authors permission to utilise a small sample of this data-base for the development of this text to demonstrate the use of national building cost data. In the preparation of this text the files have been made available via the website. These files can be imported into spreadsheet or database packages for further manipulation.

> The BCIS website address is www.bcis.co.uk

The tender price index information that is relevant to the cost analyses is given in the analysis details. The relevant locational index information for the case study project and the above analyses are also shown on the website. The methods available to develop indices have been discussed in previous chapters and will not be considered in their application.

9.2.3 Cost models

The cost models available for reader manipulation are designed to reflect current practice and should be used in conjunction with the design information available on the book's website. The assumptions made in their development have been made explicit and the reader can alter index, floor area and cost ranges to test the sensitivity of the cost bracket to such changes. The text in the next three chapters avoids the use of contemporary cost information, as it would soon become out of date. The book limits its scope to the consideration of the processes involved in the provision of cost advice. In some instances, cost information has been used to demonstrate a principle.

The reader is referred to the book's website for the most up to date supporting information, when this information is required it will be indicated as before. The information is sectionalised to highlight its relevance during the development of the cost plan, this sectionalising is purely for presentation purposes, in practice the information would be made available and analysed continually as conjectures are tested and refuted or confirmed as the design develops.

Although the text has not been written in an open learning format, the reader is encouraged to attempt the reflexive exercises and refer to the website to test his/her (and the authors') assumptions.

 Go to the website to view the information supporting this chapter.

Figure 9.1 illustrates the approach taken in the use of a case study to demonstrate the process of design cost management and how the web-based material supports the text and the differing stages of the case study's design development.

As discussed in the previous chapters, at the earliest stages of a project, the construction economist's role is to provide the client with early cost advice. Information upon the function of the proposed project is gathered from the client prior to considering its form. In the case study used to explore the design cost management processes it has been assumed that the final design has yet to be determined and that the design team have a *tabula rasa* for design development.

Chapter 7 considered the two approaches that could be taken at this early stage to establish the cost bracket. The first was to consider whether any performance indicators exist for the building function being considered. The second was dictated by the building's market value to a developer. In the first case the function would be determined via discussions with the client focused upon how the building's performance would be measured in terms of its utility in meeting business aims. In the case of a school this is tightly prescribed by the Department for Education and Employment; strict guidelines exist with regard to space per pupil for teaching, dining, administrative support and services. In an industrial context the relationship between a building's space and its related function are looser. There

Process description

Web based case study material

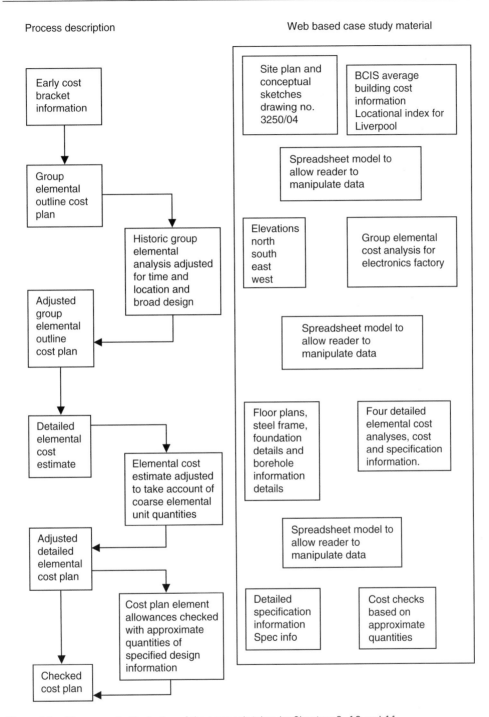

Figure 9.1 Diagrammatic illustration of the approach taken by Chapters 9, 10 and 11.

are obvious constraints regarding health and safety requirements and optimum space per worker, which would be explored with the client and design team at a later stage.

The construction economist would need to try to understand the client's business needs in as much detail as possible in order to provide the most appropriate form of advice and to search for cost data that related to the client's functional requirements.

Reader reflection

Draw up a draft agenda for your first meeting with the client to ascertain functional requirements for an industrial production and office building.

Go to the website to view a draft agenda for a typical early meeting with a client.

If the development value method were used to determine the cost bracket, the construction economist would have to be in a position to estimate the value that would be generated by the development. In the case of a school or university it would be relatively straightforward to predict income generation that may be based upon information that was reasonably stable such as student numbers and government funding. The situation in an industrial context however is a lot more complex and will involve an in-depth assessment of the production value of the asset under consideration. This will involve a detailed analysis of the cash flows related to the new facility over a period of time that constitutes the decision period. The assumptions that underpin the cash flow predictions would need to be examined in great detail and an element of risk assessment would be under-taken for each one of these assumptions. The value of the completed asset would be shown on the balance sheet and its evaluation would be determined by the accounting policy of the company and would take into account taxation writing down allowances, net present values and internal rates of return and depreciation. The methods available for business investment decision analysis are outside the scope of this text and the reader is recommended to review Baum and Crosby (1995), Enever and Isaac (1996), Darlow (1998) and Steley (2000) for a detailed discussion on investment decision-making. The role of the construction econo-mist at this stage should embrace a lot more than simple capital cost advice and a good knowledge of how a client values a development is essential. An example of a residual calculation is given below.

Example

An inner city office block has gained planning approval for an eight-storey proj-ect of 3200 m^2 total usable office space. The client's property adviser has provided the following assessments of income and yield from the proposed development.

- Income from office rents on completion
  in 18 months time   £200 per m^2 per annum
- Expected rate of return, or yield   8%
- Land purchase price   £1 525 000
- Finance rate   11%

- Allowance for risk and profit   15%
- Consultants' fees   10%
- Pre-construction period   6 months
- Construction period   12 months
- Letting period   Assume none
- Building efficiency   80%

A   Income (value)

$3200 \, m^2$ usable area at £200 per m^2 (@ 8% year's purchase in perpetuity)   £8 000 000

B   Costs

Land: purchase price	£1 525 000

Allow for financing land costs for pre-construction and construction
    period = 6 months (pre-construction) + 12 months (construction)
       = total 18 months or 1.5 years.

Effective finance = 11% $\times$ 1.5 years = 16.5% total

Allowance = 1 525 000 $\times$ 0.165	£252 000
Land cost and finance	£1 777 000

Land taxes and other acquisition expenses — *not* included

Profit and risk 15% of value of development = 15% of £8 000 000	£1 200 000
Marketing – Not included	—
Total costs (deducted from value)	£2 977 000
Gross residual construction value	£5 023 000

This sum is the budget for the whole of the proposed building works including allowances for consultants' fees and finance for the construction period. So, the total costs represent:

Building costs   100%
Consultants' fees   10%

Finance on construction period (12 months) with no letting period (all pre-committed) for 50% of construction period (standard approach)

    = 12 months = 0.5 year $\times$ 11% = 5.5%

Total   115.5%

$$\text{Building costs} = \frac{5\,023\,000}{1.155} = 4\,349\,000 \text{ say } £4\,350\,000$$

Building area = $3200 \, m^2$ (net)

$$\text{Gross floor area} \atop \text{(80\% efficiency)} = \frac{3200}{0.8} = 4000 \, m^2 \text{ GFA}$$

$$\text{Target building cost} \frac{£4\,350\,000}{4000 \, m^2} = 1087.50 \text{ per } m^2 \text{ say } 1080 \text{ per } m^2$$

> This target cost of £1080 per m^2 is inclusive of external works, car parking, contingencies, location allowances, fluctuations, time indices and any other allowances. This cost should be compared with similar projects to assess whether it is feasible to complete these offices to the cost, quality and time as indicated by the design and client teams.
>
> Obviously the determination of size or floor area of the project is precisely the same as in the first approach except the budget is determined by the market for

the property and this value is tested against the construction market price for this type of project, specification, geometry and location.

The need to spend time in identifying the client's requirements in detail has been highlighted by numerous authors of government reports. In 1964 Banwell identified that there was an urgent need to define the project requirements and that more time should be focused upon this critical stage (Banwell 1964). This recommendation was echoed by numerous reports that followed (Latham 1994, Levine 1996, Egan 1998). For a more comprehensive discussion of the processes of briefing the reader is referred to O'Reilly (1987) and Barrett (1999), which examine in detail the stages that assist the various parties in communicating their requirements to each other.

Reader reflection

Draw up a list of factors that you feel would influence the value of an electronic components production facility.

Go to the website to view the authors' suggestions.

9.2.4   *The case study brief*

A client has indicated that a manufacturing facility for electronic components is required in the north-west of England that is capable of increasing existing production capability by 25%. The organisation's current output is produced by a manufacturing facility in the south-east. However, the organisation is attracted to Liverpool due to the availability of skilled staff and its good communication links. For the purposes of this text it has been assumed that the client has a need for a new facility and has secured an option to purchase a site on a technology park. A broad capital cost figure is required by the board of directors in order to approve the release of finance.

The construction economist may be employed by one of a range of organisations at this early stage that may be determined by the procurement methods used by the client. The availability of information to inform the advice given will also vary considerably with the type of organisation. If the construction economist was working for a design and build contractor an increased amount of construction process information may be accessible from the various departments such as buying, planning and estimating when compared with a construction economist working for a consulting organisation who may have access to more information regarding development values, availability of grants and land values. This access to, and understanding of, information would have a major influence on the format and reliability of the early cost advice given.

At this stage of the project the cost economist is often given 'soft or provisional' design information on which to base the early cost advice to the design team. The reader should review the drawings and briefing information shown on the website and reflect upon the areas of uncertainty at this early stage. As stated earlier cost information is stored by the BCIS using a functional taxonomy, i.e. average building cost for factories producing electronic components are grouped together under CI/SfB coding 275.5. The average cost data are also stored in a form that reflects the technology used, i.e. steel framed, loadbearing brickwork, concrete frame or some other form of construction.

The interrogation of such databases at this early stage therefore allows both form and function to be considered. The reader should reflect upon the assumptions that have been made when deciding upon a classification system for the storage of cost information. The need for a rigorous taxonomy that is logically sound for the storage of cost information has been highlighted in Chapter 5 and the reader is referred to Bindslev (1995) for a discussion on the rationale for structured data storage. The text returns to this important area in Chapter 12.

Reader reflection

What information would you request from the client at this stage to enable you to give an outline figure? Give a rationale for why you would need this information at this stage.
What problems do you envisage in gaining the information that you require?

Go to the website to view the authors' suggestions.

The following information should be considered at this stage. The information may not be 'firm' and will be amended as the design develops. It is due to this flux in information provision that it is essential that the construction economist makes clear to the recipients of the cost advice what information was available at the time of the advice:

■ size range (in m^2)
■ time constraints, such as time to commencement of construction (this may be affected by grant availability, market entry of manufactured products, etc.), time for the construction and the dates for completion
■ location of site
■ site features, i.e. brownfield/greenfield, ground bearing capacity, ground water
■ quality indicators, such as extent of servicing, functionality of the building.

This information often becomes available at differing stages during a project's development. Some of the factors that may affect the collection of this information include:

■ client's internal organisational politics
■ consultants' differing working practices
■ time pressure
■ lack of resources
■ inexperienced clients.

The information given in the case study is a summary of the client's brief at this first stage.

Case study

Client: Cranium Containers Ltd.
Facility: Production facility for personal computer add on cards.
Business strategy: The client requires an increase in production facility and has been advised that a medium-sized building in the north-west is required.
Production requirements: Highly serviced electronic plant required with bench-style production facilities. Typical services from the plant in the south-east (refer to drawings).
Employees: 30 semi-skilled employees and 12 managers/support sales staff.
Site: Brownfield site on outskirts of Liverpool, service availability OK, access from main road.
Building: Approximately 1000 m^2, required within 18 months.
Quality: Building to act as regional centre for sales and client meetings.

Go to the website to view further information about the client.

9.3   Early cost advice

Clients often consider the initial cost advice of the building as the cost limit for the project. This is an unfortunate use of information that is unreliable, however understandable, as the client has to establish the financial feasibility of a project to gain the necessary backing by the board of directors or funding organisations. The construction economist should ensure that the client is aware of the limitations of the advice given at this early stage. The communication of early cost advice is a difficult area as the different parties giving and receiving the advice imbue it with differing meanings. The reader is referred to Bowen and Edwards (1996) for more information on this aspect of communication.

In order to provide an initial capital cost figure the construction economist has to have an idea of the overall size of the facility or the advice will have to be limited to broad cost indicators such as a range of costs per m^2 for similar projects or per functional unit, i.e. per pupil or per bed. The broad size range and function of the facility will establish the first criteria by which to search the office or national databases for relevant cost data.

Many publicly funded projects have to comply with specified funding constraints such as cost per m^2 for registered social housing or area/cost per pupil for school projects. Examples of external constraints that will have to be borne in mind at this early stage if the project was being constructed in the public sector are shown on the website.

Go to the website to review examples of external constraints on project budgets.

The BCIS average building cost data are available for new build, refurbishment or extensions to existing facilities. The average tender prices are stored as £/m^2 using a functional taxonomy under table 0 of CI/SfB. For example, the CI/SfB coding that had been allocated for factories for electronic, computers, or the like is 275.5.

Building Function	£/m^2 gross internal floor area						Sample size
	Mean	Lowest	Lower Quartile	Median	Upper Quartile	Highest	
275.5 Factories for electronics, computers, or the like	733	449	655	674	778	1148	10

Figure 9.2 Average building prices for factories for electronic computers or the like.

A cost per gross internal floor area (£/m^2) rate is available for past projects that *excludes* any allowance for external works and contingencies and have the preliminaries apportioned by cost.

The likelihood of all 10 projects within the sample above being tendered at the same time and location is extremely remote. Therefore, the BCIS adjust the tender prices submitted to a common date using the Tender Price index and a mean UK location. The Tender Price index is stated (e.g. 169) for the first quarter of 2001. The mean locational index data uses a datum of 1.00 as the UK mean. The processes of developing and applying index information have been discussed in Chapter 6.

From the data in Fig. 9.2 the construction economist can use the following information to form a judgement:

- mean tender price
- the range of tender prices of past projects
- the spread of the tender priced based upon the quartile data
- an estimation of the reliability of the data based on the sample size.

Great care has to be taken when using this information as no indication is given as to key design information such as the differing sizes or quality of the factories within the sample, the locational or tender date of the base projects or the type of construction used. As the projects within the sample have been constructed using a variety of procurement routes the data are of a coarse nature and should not be considered as being too accurate. The BCIS recognise that these as well as other factors are 'hidden' by the average building cost data and publish a range of studies that consider how the selection of a contractor, the height of a building or the contract sum influence the tender prices.

A number of larger quantity surveying organisations in recognising the problems of the reliance on professional judgement and the difficulties that the BCIS has in capturing cost information have developed richer databases of cost information. These databases contain not only the elemental breakdown of the costs but also incorporate the 'knowledge' of the construction economist as the design develops. A rationale for each decision made by the design team and the construction economist judgement is recorded. The construction economist reflects upon the relevance of each assumption made as the design develops and this reflection is also recorded and is made available to future users of the database. This 'legacy' archive also includes information about the context of the advice, the outturn costs and variations that occurred as the project progressed. Modern database technology and information and communication technology allows images, video and audio to be captured that will allow project histories to be recorded and made available extensively in future.

With the exception of the emerging supporting technologies, all of the above sources utilise tender price data rather than final account data as they are easier to collect and less sensitive to the post contract influences. The data reflect the price that clients have paid for buildings in the past and are considered to be a reflection of the national and local market.

9.4   Selection of a model

As discussed in Chapter 5, the construction economist has a number of different modelling techniques that can be used at each stage. Fortune and Lees (1996) and Bowen and Edwards (1996) highlighted the selection and use of cost models in practice. The model used primarily depends upon:

- the design information available
- the experience of the construction economist with the techniques available
- the format and extensiveness of the cost information available
- the expectations of the client/design team.

The practice of other modelling techniques is discussed on the website. At this stage of the process of design cost management the superficial traditional model is adopted.

Assuming that the design can be quantified at this stage to give a floor area, a cost rate can be applied to the design information that must reflect a number of factors:

- the quality of the design information available
- the subsequent use of the cost advice
- the experience of the construction economist
- the level of confidence in the available data.

The case study design information available at this stage is only spatial, i.e. $1000\,m^2$ and may have been derived from the designer's knowledge of the requirements of similar facilities. The client has provided a broad idea of the function of the facility and through questioning the client about other production facilities an impression can be gained regarding the performance of the facility.

The client's expectations regarding the timing of the completion of the project need to be explored as this could have a radical effect upon the budget estimate. If completion is expected very early a procurement route that required the constructor to manage a large proportion of the risk of design and construction from an incomplete brief may lead to a premium being paid. If the client is inexperienced advice may be needed regarding the likely timing of the various stages of the pre-construction stages of the various procurement routes. The projected tender period date would require discussion with the client to allow for price trends in the market to be taken into account, the case study has assumed a period of three quarters, i.e. 9 months, between clients briefing and construction commencement. Unfortunately, little is known regarding the causal effects of speed of construction and procurement method upon construction cost. The construction economist would have to search through similar historic projects to gain an impression about whether the client's time periods for the various processes of briefing, design and construction are reasonably achievable within the procurement routes available. Newer procurement systems such as partnering aim to bring the constructing parties together at an earlier stage than the traditional systems. These newer techniques provide opportunities to make use of more informed judgements from a broader range of cost information sources regarding time and technologies at an earlier stage. The impact that these newer arrangements will have upon the processes of design cost management will be discussed in Chapter 12.

To illustrate the difficulties of using cost data without being aware of their variability, Fig. 9.3 shows rates for factories taken from the BCIS. We need to gather more information from the design team to allow us to refine our budget estimate. The construction economist has to formulate the advice that is given to the client, this advice is a synthesis of cost data, design information and judgement and could be given in the following format:

Client: Cranium Containers Ltd
Date: 06/12/01
Building function: Electronics factory
New build
Location: Wavertree Technology Park, Liverpool, UK
Tender date: 2nd quarter 2002

Information available
Client discussion at meeting dated:
Engineers sketch plans ref. 23648/p01, 02, 03
Gross floor area: approximately $1000\,m^2$, based on engineer's drawings

Cost bracket:
Floor area: $1000\,m^2$

Adjustment for Price Risk
Average building price rate range $£/m^2$ GFA $\times$ 1 + Difference in Tender Price index at projected tender date and current Tender Price index/current Tender Price index.

Adjustment for location
This gives a range of projected tender prices taken to the date of tender for the project. The range is to a UK mean location and still requires adjustment for geographical location.
Average building price rate range at UK mean location $\times$ locational index.

Adjustment for Design Risk
The average building price excludes any allowances for external works and contingencies which must be included when arriving at the project cost bracket. The construction economist must arrive at a judgement regarding the capability of the design team, the quality required by the client and the reliability of the sample of cost information and make an appropriate allowance.

All the figures used in the case study exclude VAT. Most building costs are subject to the addition of VAT, however, this may alter if the client's status is a charity or if the building is listed. The reader is referred to the Inland Revenue website for more information.

Rate per m² gross internal floor area for the building excluding external works and contingencies and with preliminaries apportioned by cost. Last Updated 26-Jan-2002.

At 3Q2001 prices (based on a Tender Price Index of 186) and UK mean location.

Building Function		£/m² gross internal floor area						
		Mean	Lowest	Lower Quartile	Median	Upper Quartile	Highest	Sample
	New build							
275.5	**Factories for electronics, computers, or the like**	829	529	704	785	837	1354	10
282.	**Factories**							
	Generally	475	149	329	413	555	1970	832
	Steel framed	461	149	321	398	542	1970	698
	Concrete framed	662	246	429	585	825	1462	27
	Brick construction	512	261	387	447	602	1115	105
	Timber framed	395	–	–	–	–	–	1
	Up to 500m² GFA	580	292	448	519	650	1630	120
	500 to 2000m² GFA	475	190	340	415	549	1970	370
	Over 2000m² GFA	437	149	297	370	515	1527	342
282.12	**Advance Factories/ Offices – mixed facilities (class B1)**							
	Generally	577	265	381	578	700	1171	89
	Steel framed	559	265	361	548	677	1171	68
	Concrete framed	642	437	537	585	765	961	9
	Brick construction	632	387	543	631	706	1042	12
	Up to 500m² GFA	581	387	441	602	700	778	5
	500 to 2000m² GFA	617	275	435	592	710	1132	36
	Over 2000m² GFA	548	265	374	530	681	1171	48
282.22	**Purpose built factories/ Offices – mixed facilities**	548	242	422	523	619	1214	49

Figure 9.3 Average building prices: factories.

Go to the website to view the authors' suggested cost bracket.

The web model has taken a range of average costs that are higher than mean costs per m² due to the potentially high servicing element of the project. This service element and the need for anti-static floors used in electronics production facilities alert the construction economist to potential deviations from the functional mean cost. The specifications for similar projects could be reviewed relatively quickly on line to gain an impression of what is typical and atypical for a project of this nature. The early cost advice bracket has been adjusted to take into account the estimated increase in tender prices that the BCIS judge will occur between the date the advice was formulated and the projected date of tender or agreement of constructor price. This judgement is based on national macro economic data; it does not take account of regional differences that are taken into account by the regional locational indices. The locational adjustment for constructing in Liverpool in the north-west of England is 0.98, i.e. the Liverpool market price is 9% lower than the national mean. An allowance needs to be added for external works as these are excluded from the average building price data due to reasons discussed in Chapter 5. The case study is situated on a brownfield site on a new technology park. The construction economist would visit the site as early as possible to gain an impression of the cost-sensitive features of the site; these would include topography, location of existing roads, sewers and electricity and gas services. The services are near to the proposed site and a judgement has to be made upon the extent of the external works. In the case study the client has given us an estimation of the number of employees and we can consequently make a working assumption upon the number of car parking spaces required and the consequent more expensive hard landscaped space required.

The resultant cost bracket figures need to be qualified before communication to the design team or client. The cost bracket shown above could be further refined to take account of further adjustment factors that have been highlighted by recent research by the BCIS.

Fact file

A sample of over 6000 projects whose tenders were based on bills of quantities were indexed as part of the process of compiling the Tender Price indices and the following factors were investigated:

- date
- location
- regional trend
- procurement route
- contract sum
- building function
- building height
- type of work.

The analysis demonstrated that all these factors were statistically significant and a further set of tables have been compiled by the BCIS that can be used for finer adjustment produced for:

- location
- contract sum
- procurement route
- type of work
- building height
- building type.

Go to the website to view the factors that apply to the case study and consider their utility at this stage.

9.5   Summary

The purpose of this form of cost modelling is primarily to give to the client and the design team an estimation of the probable cost of construction. The model used by the website is referred to as superficial in that it uses a project's superficial floor area as a basis for development of advice. It is popular with practitioners and the resultant advice is easy to communicate to all parties. The flaws in this traditional modelling process are:

- The quality of the design data. In the above scenario the figure of $1000\,m^2$ is obviously a very rough estimate and may change radically as the design team interprets the client's requirements. The model does not reflect this potential variability and little information is available to the construction economist to assess whether the size of a facility is within an expected range.
- The source and reliability of the cost data that are applied to the early design information. The construction economist must select a range of rates from the cost information available to apply to the design information. The reliability of the information available has to be taken into account when selecting a rate range that is appropriate. Recent past experience of similar project types that have been tendered and constructed would obviously take priority over published data that was of unknown provenance. The BCIS have recognised the difficulties of making decisions on historic data and publish average building prices but also statistical information giving the range, mode, standard deviation and quartile information as discussed in Chapter 5. The BCIS data can be referred to as data from a secondary source and is used in the above case study without any idea as to the quality of design or variability in quality of the projects that make up the sample.

The reader is referred to the website model to simulate a number of scenarios for the project; the model allows for a change of tender date, floor area and regional location index. Note how relatively small changes to this information can have a large impact upon the range of early cost advice given to the client. Chapter 10 will consider how the cost brackets are narrowed to provide more reliable and useful information in order to assist in the cost management of the design process.

9.6   Reader reflections

- Consider the range of average building costs for electronic factories shown in Fig. 9.3, how would you assess the reliability of these data?
- Review the conceptual drawings on the website. What areas of design and construction risk would you identify at this early stage? Consider what information you would require to start to analyse the risks at the design stage and what information would be available.
- How would the client's use of the early cost advice affect the format and extent of advice given?
- What do you understand by the term qualification? Should your advice be qualified and, if so, how would you communicate it to a client and design team?

10 Refining the cost advice: sketch plan stage

10.1 Introduction

This chapter aims to:

- Introduce the reader to the processes used to arrive at a detailed elemental cost plan
- Demonstrate how a cost analysis is adjusted for Price and Design Risk
- Give examples of typical design information that becomes available at the sketch plan stage
- Show how an elemental cost target is developed
- Discuss the sources of error within the cost data that are due to procurement practice and construction technologies.

The upper bracket of the early cost information will form the cost limit for the project. However, the design team will need more specific advice regarding the cost consequences of the design of the project at a finer level. The construction economist can provide the design team with a target (or standard) to work to for each of the functional elements that make up the design. These standards are generally derived from historic cost analyses of previous projects of a similar function.

10.2 Planning the cost advice process

The management of the production of cost advice is primarily influenced by the amount of resources available to the design team. These resources include the availability of design expertise, the time available prior to construction commencing on site, the extent of the client's brief and the experience of the client. If the client has a very well defined brief that requires little design development and also has detailed design information available the process described in Chapters 6 and 7 may well not be followed as it is redundant. The timing of the flows of design information to the construction economist and consequent cost advice varies significantly from project to project. The RIBA plan of work discussed in Chapter 5 can be thought of as a generalised model that may be adapted for each project to take account of the specific project variables. For example, the detailed design information for the substructure of a project may well be available and used for cost checking the substructure elemental target well in advance of the mechanical and electrical element cost target being produced.

In another example illustrating variance from the above, a repeat supermarket builder in the UK has used a particular group of contractors to construct a virtually identical design in a number of locations in the north-west. One contractor in the north-west has stated it only takes 4 hours to provide a lump sum, firm price tender as the majority of the costs are agreed with the supply chain subcontractors prior to receiving the tender enquiry.

The source of design information upon which to base cost targets may also vary between projects and often relates to the design team's expertise. The use of specialised contractors has led to a change in the approach architects and engineers use to design projects. Often a specialised contractor will be involved early in the design process and will provide cost advice for particular elements of the work. If the design team's intention is to have the work completed by the specialist on site the contractor may provide cost checking advice as the design develops. This issue will be discussed in more detail in the cost target section and Chapter 11.

10.2.1   Case study

It is assumed that the design team have now produced elevations and some outline floor plans that can be used to start to develop a more accurate assessment of the price the client may pay for the project.

Figure 10.1 illustrates the use of the web-based information and cost models to produce the cost plans.

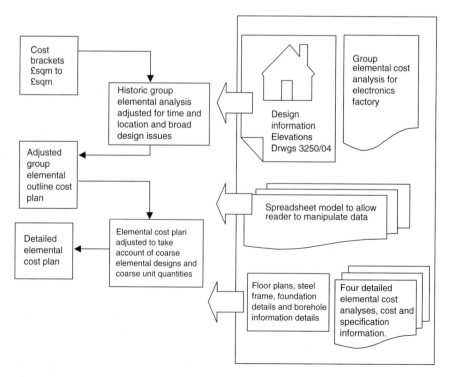

Figure 10.1   Processes and information involved in the development of a detailed cost plan from outline proposal to scheme design stage.

10.3   Development of the design and the provision of cost advice

The reader is referred to the following drawings on the website and cost information listed in Appendix 1 to gain an impression of the information that is now available from the design team:

- site layout (drawing no. 3250/04)
- south elevation
- east elevation
- north elevation
- west elevation.

Log on to the website to view the drawings.

The design team may be starting to firm up ideas regarding the optimum use of the site and the construction economist would have a major input at this stage. Information about the timescale and costs of the various elements would be sought as various technological solutions to design problems are considered.

This text is, by necessity, written in a linear fashion however, in practice this system of conjecture, reflection and possible refutation of the design would occur many times. The skill of providing an input to these early design development processes is developed over a period of time in practice and involves an intimate and interdependent knowledge of procurement practice, technological solutions and applied project economics. The text will explore some of these issues with regard to the technologies available for a building of this type later in this chapter. This exploration cannot be comprehensive and is aimed at provoking reflection on the complexity of the cognitive processes that underpin the provision of useful design team and client advice.

The sketches on the website can be considered as an early output of the design process and can be used as a basis for discussion with the client and the rest of the design team. As can be seen from the drawings, a two-storey building is suggested with a curved low pitch roof and extensive glazed area. The sketches indicate an extent of external brickwork and some preliminary ideas regarding access.

The cost models that are used at this stage reflect the availability of the design information, i.e. general and sketchy, and the advice given to the client and design team should reflect this input. The BCIS group elemental format will be used to provide the client with an outline cost plan. As we are still at the early stages of the design development, the cost advice we should give should reflect the quality of information that we have. The reader is referred to the website to view a number of typical group elemental cost analyses.

Log on to the website to view the elemental cost analyses.

The format of a group elemental cost analysis is shown in Fig. 10.2.

In order to produce the group elemental outline cost plan for the case study, a historic cost analysis has been selected and the costs per m^2 gross floor area were updated for time and location and then applied to the gross floor area of the case study. The resultant figure was then apportioned to the various elements in the same proportion as the original analysis, i.e. if substructure made up 4% of the original

General information	
BCIS Function:	275.5
Type of work:	New build
Gross floor area:	2000 m^2
Job Title:	Albert Dock development
Location:	Liverpool
Dates:	September 1999

Project details	
Market conditions:	Keen
Client and Tender documentation:	Private, Bills of Quantities
Contract breakdown:	*A breakdown of the contract may be given here*
Brief design details:	*A brief description of the specification, i.e. number of storeys, servicing and quality indicators.*

Cost information

	Elements Total cost £	Cost per m^2– GFA £
1. Substructure	120 000	150.00
2. Superstructure	245 890	307.36
3. Internal finishes	19 750	24.69
4. Fittings	7 500	9.38
5. Services	350 560	438.20
6. External works	40 500	50.63
7. Preliminaries	85 000	106.25
8. Contingencies	10 000	12.50

Figure 10.2 Format of a group elemental cost analysis.

analysis, this percentage would be allocated to the case study project. There are dangers in applying this principle unthinkingly. For example, if the base analysis has a large proportion of fittings or a high quality of internal finishes and the case study has no fittings and low quality finishes an over-allocation of cost could be made to these areas.

10.4 Searching and selection of a historic cost analysis

The first activity that the construction economist has to do is find some relevant cost information to inform the advice to be given. In the case study, a group elemental breakdown of a previous project that could be used as a basis for the development of an outline cost plan is sought from the BCIS. The various sources of cost information have been discussed earlier in this chapter and in Chapter 5.

The BCIS on line system can be considered as a large database that can be searched by a number of predefined fields. The user, after logging on, can search through the database using a number of different criteria. These criteria have been predetermined to make the searching as effective as possible.

> The BCIS website address for subscribers only is:
> http://service.bcis.co.uk

10.4.1 Searching criteria

Building function: This is related to the CI/SfB coding system. Care should be taken not to be too specific regarding the building function as function and cost are not always linked and the search may preclude some suitable source projects.

Type of analysis: Group elemental, detailed elemental.

Type of work: New build, refurbishment, extension.

Number of storeys: One, two.

Gross floor area: A floor area range should be inputted, as the search engine will match analyses with exact criteria, for the given example a reasonable range should be used. The gross floor area of the project was roughly 1000 m^2, therefore a search range of 500–2000 m^2 could be used to find analyses that are of a similar size.

Location: A code can be inserted to limit the search to the north-west of England if required. However, care should be taken not to be too specific.

Date: This is the date of tender of the historic analysis. As with other field criteria a range should be used to ensure that analyses are returned after the search has been undertaken.

After the criteria have been entered the search returns a number of analyses that meet the inputted criteria. If the ranges are too narrow and tightly prescribed very few analyses would be returned. Alternatively, if too broadly defined the construction economist could be overwhelmed with information and need to redefine the criteria.

The following criteria were used to search the BCIS on-line database:

Building function:	275.5 — Factories for electronics, computers or the like
Type of analysis:	Group elemental
Type of work:	New build
Gross floor area:	500–10 000 m^2
Location:	North-west UK
Date:	1993–2000

Log on to the website to view the group cost analyses using the selection criteria noted above.

One group elemental analysis was found from the search and is shown below. The full specification and project details can be viewed on the website. The use of a single analysis at this stage could cause errors if the design detail was not representative and in practice a number of analyses would be reviewed before selecting the most appropriate analysis. This approach is discussed later in the chapter.

> Production facility for electronics, factory A, new build, GFA 5450 m^2, 1(2) storey, September 1999, total cost £2 892 341, cost per m^2 £530.7, group elemental analysis.

The group elemental analysis could be used in one of two ways at this stage:

(1) The product of the gross floor area and a selected figure from the previously calculated cost bracket could be redistributed in the same proportions as the

selected group elemental analysis, i.e. the substructure element of the selected analysis forms 4% of the overall sum. This percentage is applied to the product and forms the updated group elemental total.

(2) A selected group elemental analysis' elemental totals are adjusted for time and location and the cost per m^2 is then applied to the gross floor area of the project.

The first method uses the original cost bracket information as a basis for further advice and whilst this may be expedient it may lead to errors. The second method uses finer cost information and, assuming that the selected analysis is reasonably similar to the design of the proposed project, would provide more reliable cost advice.

Log on to the website to view the case study group elemental outline cost plan.

This format of cost advice is still in a very coarse form and the construction economist needs to start to gather further design information at these early stages to start to increase the accuracy of prediction and develop well-informed cost targets. A good source of information on the relative different costs of design elements for buildings of a similar function is the detailed elemental cost analyses published by the BCIS.

10.5   Preparing the detailed elemental cost plan

The construction economist has to have a detailed knowledge of the technology and how it is applied to a design solution in order to provide informed advice to the design team. The design brief at this stage may or may not be comprehensive. However, in the case study major decisions have already been taken regarding the design solution. The reader is referred to the website to study the design brief in more detail.

Go to the website to view the following design information that is available at this stage and shown on the drawings identified. These can be viewed on the website.

- Floor plans   drwg no. 3250/11
- Foundation proposal drwg no. 23648/p100
- Floor layout   drwg no. 23648/p101
- Roof layout   drwg no. 23648/p102.

The design decisions that have been taken need to be considered and carefully analysed under the following categories:

- irreversible
- interdependent
- independent
- firm
- variable.

The extent of control that the construction economist has upon the cost consequences of design decisions can also be categorised in a similar manner. For example, the decision to adopt a steel frame on pad foundations would in all likelihood close down any discussions on strip footings with load-bearing brickwork technologies.

Reader reflection

Review the design information and list the decisions that have been taken already that will have a major impact upon the likely costs of the proposed building. Categorise these decisions and suggest some reasons why they may change.

The BCIS database is searched again changing the criteria from group elemental to detailed elemental.

A further four matching analyses have been found as a result of this revised search. They are all factories:

(1) Factory B, industrial estate, new build, GFA 1850 m^2, 1 storey, July 1997, total cost £1 319 301, cost per m^2 £713.14, full elemental analysis.
(2) Electronics factory C, new build, GFA 4790 m^2, 1 (2) storey, May 1996, total cost £2 479 873, cost per m^2 £517.72, full elemental analysis.
(3) Electronics factory D, new build, GFA 2865 m^2, 1 storey, April 1995, total cost £1 105 892, cost per m^2 £386, full elemental analysis.
(4) Electronics factory E, GFA 7100 m^2, 1 (2) storey, March 1993, total cost £2 237 353, cost per m^2 £315.12, full elemental analysis.

The specification and contract details of these analyses can be found on the website.

Log on to the website to view the project details.

The construction economist now has to use his judgement to select an appropriate analysis for use as the base analysis. This is a difficult decision and one that is fraught with danger. Consider the above analyses:

■ Very few are of a similar size to the case study project.
■ The range of costs is very broad from £315/m^2 to £713.14/m^2. These would need to be re-based to give us a better idea of the variability.
■ A large number of projects are based in Northern Ireland; the economic reasons for this location are probably due to grant availability. This might skew the market in this locality and give an unrepresentative indication of price.
■ Three of the projects are over 5 years old. The older the project the greater the potential error when updating using the index.
■ An early budget figure was given to the client at the early stages of project feasibility. This would undoubtedly influence the advice the construction economist would give. The construction economist should be aware of this potential source of bias that may be caused by the need to retain credibility rather than providing objective advice!

10.6 Information availability

10.6.1 Design processes

At this relatively early stage in the design decision-making process the construction economist would provide an input to the fundamental design issues. Some of these have been discussed earlier; the consequent impact of a simple decision to choose a technological solution to a design problem can be far-reaching. The adjustment for the brickwork external walling in lieu of cladding is a good example of such a decision. The construction economist should highlight these issues at the early meetings to ensure that informed decisions are being taken. If the design development is being carried out under a design and build procurement arrangement and the construction economist was employed by a contracting organisation, there would be access to a richer source of information. For example, the current situation in the UK is that there is shortage of skilled craftsmen, particularly bricklayers. This has led to quality control problems and also an increase in labour costs for brickwork. A construction economist in a contracting organisation would be more informed about prices and the local market for resources and, therefore, could provide more accurate information to the design team.

The construction economist has to draw upon an extensive knowledge of construction technologies, the construction process and procurement systems in order to provide relevant advice to the design team. Accurate and reliable cost information is often difficult to find as the procurement systems used in the UK provide a number of barriers to its storage and dissemination. As a consequence the standards and targets that are produced for the design team are often unreliable. This feature will be discussed in more detail in Chapter 12.

If one uses a base analysis that has a specification that is unlike the proposed project *over or under allowances* can be incorporated. For example, the cost analysis for factory A used for the development of the outline cost plan has extensive enabling works allowed for within the substructure. This allowance is not required on the case study project as it is to be constructed in a technology park. This illustrates the need to review all details of a base analysis before it is used.

 Log on to the website to view details.

Reader reflection

Review the cost analysis for factory A and list the elemental features that could lead to inaccurate allowances being made in the group elemental cost plan.

10.6.2 Development of the detailed elemental cost targets

The worked example on the website illustrates how a detailed cost plan can be developed from a historic base analysis. The base analysis is an electronics factory in Northern Ireland constructed in 1995. It has been assumed that the basic layout information given on the drawings can be used to measure approximate gross internal floor area and are to be read in conjunction with the outline specification information given earlier. This can form the basis for the development of the elemental cost targets. The readers must ensure that they do not scale directly from the drawings, but use the dimensions shown on the drawings.

The construction economist can use the detailed cost analyses in a number of ways to develop the basis for the project's detailed cost plan. A single analysis can be used as the basis for the detailed cost plan and adjustments made to each element as design information is made available. This method retains the integrity of the original cost analysis and the adjustments for time and location are straightforward. A number of analyses may be used, all rebased to the same date and location and then averaged. This would produce a cost analysis which, although based upon more data than the first method, would be logically unsound as the specifications of the various elements would differ. The average between a piled foundation and a strip footing is not a design commonly encountered. The exercise is worth carrying out, however, to provide background information upon which to inform judgements about the costs targets. As the cost analyses are based upon tender information such as bills of quantities or design and build contract sum analyses, the elements may have an unrepresentative allocation of price data due to contractor unbalancing or redistribution of method related or preliminary costs. A review of the preliminaries across a number of analyses would provide an indication whether the allowance made within the base analysis is representative.

The construction economist has to develop a set of targets for the design team that are used as a basis for the developing cost plan. Each element of the cost plan needs to be considered taking into account the element quality/quantity in the original cost analysis and considering the design information available for the case study project. The reader is referred to the average element costs that have been calculated from the BCIS data to become familiar with the spread of costs for a project of this type.

The cost plan should clearly identify what is included in each element, to ensure commonality between projects. The Standard Form of Cost Analysis includes a range of definitions in Section 3. For example, the substructure definition removes any uncertainty regarding what technologies and costs are to be included in this element.

 Log on to the website to view details of the SFCA.

---

1. Substructure

All work below underside of screed or where no screed exists to the underside of lowest floor finish including damp proof membrane, together with relevant excavations and foundations.

This is further expanded with a set of notes, e.g. note 3 states that the cost of piling and driving shall be shown separately stating system, number and average length of pile.

This developing cost plan is really a set of working targets that are used as a basis for the preparation of the detailed cost plan. This will be used to allocate the design team's expenditure to the various elements. The use of the cost plan as a dynamic planning tool will be discussed as the cost plan is developed. In practice much discussion will take place regarding design decisions, the impact they will

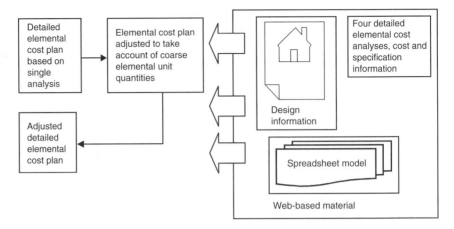

Figure 10.3 The establishment of the cost targets.

have upon the functionality of the client's building, design aesthetics and their cost implications.

Figure 10.3 illustrates how the web-based case study material supports the process of adjusting the detailed elemental cost plan.

10.6.3   Case study design information availability

The information that can now be considered as available to the design team consists of:

- site layout   drwg no. 3250/04
- south elevation
- east elevation
- north elevation
- west elevation
- floor plans   drwg no. 3250/11
- foundation proposal   drwg no. 23648/p100
- floor layout   drwg no. 23648/p101
- roof layout   drwg no. 23648/p102
- site investigation report
  and borehole information.

Log on to the website to view details.

In order to arrive at a set of cost targets that are of more use than the simple elemental costs per m^2 of floor area, the construction economist must quantify the element unit quantities at a level that relates to the element itself. The elemental allowance can be broken down into the following costs:

- cost per m^2 of gross floor area
- costs per elemental unit quantity
- costs expressed as a percentage of total cost.

Preliminary costs can be apportioned to the elemental costs, if required. However, common practice is to deal with this element as a distinctive element.

The construction economist must take account of Price Risk and Design Risk when establishing the elemental targets. The identification of Price Risks is based on a prediction of the national and local market conditions that will exist at the time of agreeing a price with the constructor. The constructor's price will be influenced by the market demand for the services offered and the price of resources required to construct the project. As illustrated in Chapter 6, the construction market does not always follow the retail price trends and the construction economist must have a good grasp of the future market conditions. This knowledge is not only gained from national sources of cost information such as the BCIS Tender Price index projections, but also RICS economic briefings and specialised sources of construction market intelligence. Often local market conditions will display an abnormal trend, for example in Liverpool at the time of publication, there has been a reduction in the number of medium-sized contractors i.e. with a turnover of between £10 million and £40 million which, when combined with the general trend for contractors to have a central northern rather than a number of regional offices, is having an impact upon the supply of local contractors with the capability to undertake a project of the case study's size. This trend may also lead to contractors being unaware of the availability of the best value resources in the project's location and reliance on known subcontractors operating in the head office region. The additional costs of transportation would have to be included in tenders and may also lead to premium costs being paid.

The case study utilises the tender price indices and locational factors from the BCIS when adjusting elemental cost allowances from previous projects. As discussed earlier, these are derived from bills of quantities submitted to the BCIS and analysed to reflect the tender price movement over time and also regional differences. As discussed above, the indices are only an indicator of price movement; they should inform professional judgement and not be used blindly.

The allowance that needs to be made for Design Risk is a lot more subjective and has to take account of the ability and previous experience of the design team and the cost drivers for each of the elements. These will vary significantly with the technologies and processes used to construct the various elements. The construction economist must have a good knowledge of the relative costs of the technologies available. This understanding is developed through consideration of elemental costs for past projects and potential source of cost information error, the technologies used and the procurement practices used by clients and constructors. The next section of this chapter discusses each of these interrelated issues by considering each element of the detailed cost plan. The elemental unit quantity adjustment will often take account of the increase or decrease in an element that may not be taken into account when adjusting for gross floor area. For example, if an analysis was used as the basis for a cost plan that had $400\,m^2$ of upper floor, this could be increased to take account of the proposed project's upper floor elemental unit quantity of $500\,m^2$. The more difficult adjustment would be for specification differences between the base analysis and proposed project. This is an area where professional judgement is used to arrive at a percentage adjustment. The reader should log on to the website to review the assumptions made by the authors for Price and Design Risk and consider the validity of such assumptions.

Log on to the website to view details of the elemental target establishment.

10.7   Establishing the cost target

10.7.1   1 Substructure

Elemental unit quantity: Area of substructure.

Sources of error within the cost data

The substructure element is generally carried out at the commencement of a project and it tends to have a large number of variations to its design as work proceeds on site. It is due to these reasons that the price data that the bill of quantities captures under the traditional procurement route may be unreliable. Contractors may have also unbalanced their tender to enhance a project's cash flow in the early stages of a project and also to inflate any profits that may accrue if variations occur in the substructure and are evaluated at bill of quantities rates.

Supply chain procurement

Few contractors carry out their own groundworks work elements. These are often sublet to labour and plant subcontractors. The relatively low barriers to entry for subcontractors for small groundworks means that the market price is relatively stable, as competition is fierce. The main contractor provides materials and also supervises the setting-out of this element of the works. The setting-out costs tend to be included in the Preliminaries element of the contractor's price. If the project under consideration has complex groundworks across a number of locations this may mean that the contractor would have to increase the supervision element in the tender and the construction economist should make an allowance in the costs of this element.

Technology

The main determinants of the costs of substructure are:

- prevailing ground conditions, e.g. rock, ground water
- intended load from the proposed projects and consequent substructure design
- extent of material that requires to be remediated or taken from site
- any constraints/requirements for construction methods that may require extensive enabling works.

In the first two instances these are design issues that would be taken into account by a structural or civil engineer. The consulting engineers would carry out a site investigation to consider the bearing capacity of the ground. This is usually carried out by digging a series of trial pits in a regular grid up to 2 m in depth. The pits would reveal the underlying ground conditions and would discover issues that would need to be taken into account when designing the building's foundations.

Cost-sensitive information at this stage would include:

- extent of rock
- extent of 'soft' spots that would require excavating and filling with suitable material
- presence of running sand/or material that is unsuitable for bearing
- ground water level
- extent of any ground contamination
- whether any existing foundations are present.

The information derived from the trial pits would be augmented with a deeper investigation using boreholes. These may be up to 20 m deep.

The project's substructure cost drivers are dependent on the load being transferred to the ground and the methods of load transfer. It is beyond the scope of this text to consider the numerous foundation solutions available to the engineer. As a guide the construction economist must keep in mind the following:

- Extent of reduced level dig or fill required. Although the cost per m^3 can be relatively low, often the volume required can lead to high costs. The key design information to be derived from the drawings would be finished floor levels, reduced levels and existing ground levels.
- Extent of specialised works such as vibro-compaction (this is a technique that compacts the ground with columns of stone to provide a more stable sub base), whether any piling is required (this would require a specialist subcontractor and may have an impact on the procurement methods selected), whether any de-watering is required during the construction phase (key design information in this case would be existing ground water levels).
- Foundation design, large reinforced concrete bases may require deep excavation that necessitates specialist construction techniques whereas simple mass fill strip footings are relatively quick and simple to construct.

The construction economist would draw upon his/her experience of the construction process at this stage. Often the contractor would require extensive enabling works on the site to bring plant into position. The plant requirements may dictate that temporary roads are required, which could be costly and may be either allocated within the substructure or the preliminaries element of the cost plan.

 Log on to the website to view details of a site investigation report and a number of borehole logs.

An example of the format of a substructure detailed cost target is given in Fig. 10.4.

The reader should develop a set of elemental unit quantities and calculate the elemental unit rates from the cost plan shown on the website. These can now be compared with any elemental unit rates for the detailed analyses that are on the website to get a feeling for their reliability. Unfortunately, not all construction economists break down the design details of their project to an elemental

Job No. 123 Cranium containers	Client: Cranium Containers Ltd	Cost plan no. 2	Date		
Element title	Substructure				
Element no.	1A				
Preliminaries apportioned	No				
Item	Description	Element unit quantity	Unit	Elemental unit rate £	Elemental unit total £
1A Substructure	Specification detail if any	528	m²	50.89	26 870
				Total to elemental summary	26 870

Figure 10.4   Format for substructure cost target.

unit quantity level, which limits this approach for comparison. If the functional analyses found following the BCIS search do not go into the level of detail required, the construction economist can browse through analyses that have a similar specification to gain an impression of the range of element unit quantity rates. These should, of course, be updated for time and location to make the exercise meaningful.

Log on to the website to view the detailed cost target build ups and the authors' rationale for the target development.

10.7.2   2 Superstructure

2A Frame

Extent: Load-bearing framework of concrete, steel or timber. Main floors and roof beams, ties and roof trusses of framed buildings. Casings to stanchions and beams for structural or protective purposes.

Elemental unit quantity: Area of floor that relates to the frame. In the case study approach this would be gross floor area.

Sources of error within the cost data

The procurement of steel and pre-cast concrete frames tends to be via specialised contractors who are often integrated into the design team at an early stage. Most analyses stored by the BCIS are for projects procured using bills of quantities; these would generally include a prime cost sum for the frame. The prime cost sum would be included by the design team and may be based on a budget given by the steelwork subcontractor. The prime cost sum may not reflect the actual cost as design team contingencies or subcontractor allowances for contingencies to cover design and construction risk may be included. This approach is reflected by the standard method of measurements approach to categorise the prime cost sum as either defined or undefined, however this information rarely finds its way into the cost analysis.

Wherever possible select a cost analysis or range of analyses with a similar number of storeys and framing system. An important decision is to identify whether load-bearing brickwork or a steel or concrete frame is needed or is likely to be used. Then follow cost analyses in the BCIS with a similar structural design approach. Also ensure that costs of floor and roof beams are included in the frame costs and not the upper floors element.

Supply chain procurement

A specialist contractor who may have the capability to complete the design and may be able to provide important construction process information to the rest of the design team at the early design stage would carry out the frame fabrication and erection. The design team often has the opportunity to bring a preferred subcontractor in at an early stage with the intention of nominating the subcontractor at a later stage. The advantage of this approach is that the risk associated with a large element of the work can be reduced considerably at an early stage and the design team would request a firm cost estimate from the specialist. The consequent construction management issues of the approach of using specialist subcontractors at the early design stages are beyond the scope of this book. However, as a general principle the construction economist needs to be aware of the extent of risk being carried by the specialist, the information they have been provided with that forms the basis of their estimations and the qualifications and conditions they make when returning their tender. The steel frame is usually carried out by a subcontractor engaged by the contractor as either a domestic or as a nominated subcontractor. If the subcontract includes an element of design then the subcontract has to take this into account. For steel frame construction, the materials used form the major costs for this element due to the weight of the steel used. The labour and plant involved in fabrication and erection is comparably small. It is difficult to gather detailed costs for these resources as the subcontractor invariably submits an all-in cost per tonne. Often the subcontractor will require specialist attendances (requirements) to be provided by the contractor that may need to be considered by the design team at an early stage. For example, a stable sub-base may be required to allow for cranes and access platforms to be used.

Technology

The client has requested a building with a relatively large amount of free space and the design team has considered that a structural frame would be an appropriate solution. In arriving at this solution the design team would have considered:

- length of span required between structural elements, the larger the span the more flexible the space utilisation
- speed of construction
- potential cost of load-bearing external and internal walls.

A framed solution can be quick to construct, as a great deal of the construction work takes place off-site in a steel fabrication yard or in a pre-cast concrete manufacturing facility. Standard sections are used which allow for savings to be achieved through the casting or rolling process. *Walmart,* a major client who procure supermarkets, cited an example of using the supply chain to gain savings. As a significant client they influence the design of their stores by prescribing the section sizes that are to be used by engineers when designing the steel frame. These sections are bought in bulk from the rolling mills and supplied to the steelwork subcontractors for fabrication and erection. As they build numerous stores a year the consequent material savings are significant.

The design team has to consider which type of material to use, either steel or concrete. Steel is generally cheaper but requires costly fire protection. The extent of fire protection is generally dictated by the use of the building and can vary from a simple boarding of a *supalux* material to an extensive *in-situ* concrete cladding.

A structural engineer would have a computerised design system that will enable a number of frame solutions to be considered. Some of the issues to be considered are:

- Span: the longer the span the greater the section size and consequent weight that is required to be carried by the columns.
- Loading, e.g. snow loads, services loads, air-conditioning units to the roof will increase dead loads and would have to be taken into account. This interdependence of the design with the various elements will be discussed later in the chapter. The structural engineer will need to make explicit the assumptions of loading made when designing the frame as the cost consequences of not considering other element interactions could be disastrous.
- Height of columns.
- Number of storeys.
- Type of floor construction/use of building.

10.7.3   2B Upper floors

Extent: Upper floors, continuous access floors, balconies and structural screeds … suspended floors over or in basements stated separately.

Elemental unit quantity: Area of upper floor.

Sources of error within the cost data

The project upper floor cost data are often based on a subcontractor's quotation as this work is often specialised. The quotation will invariably be qualified in some way regarding access or temporary works provision. The costs to the constructor include an allowance for these qualifications and may be included within another section of the tender document, such as the Preliminaries. If the contractor is carrying out the works for example, an *in-situ* concrete floor that utilised permanent formwork, then the costs of transporting the concrete into position (such as with a concrete pump) may be allowed within the preliminary section and remain hidden from the construction economist.

Supply chain procurement

The decision to subcontract the works is dependent upon the technology the design team adopt. A simple timber floor or *in-situ* concrete on permanent formwork would generally be completed by constructor's labour. If the work was more complex or had a design element such as pre-cast concrete floors the contractor may sublet the design and supply of this section of the works to a specialist subcontractor. The subcontractor may require certain attendance such as access platforms and may exclude certain aspects of the element such as structural screeds that the contractor would have to allow for in the tender.

Technology

The design of upper floors can be a very cost sensitive aspect of a project. A range of design solutions are available. Domestic construction tends to rely on softwood joists and floorboards. Industrial solutions also include reinforced *in-situ* concrete and either self supporting or poured on permanent formwork such as *Holorib* or a pre-cast concrete floor, which can be either hollow or solid. If the upper floor is to be a pre-cast solution a specialist subcontractor often completes this element. Screeding is commonly excluded and has to be carried out by the contractor if it is not a design requirement. The suppliers of pre-cast concrete floor solutions are often nominated by the design team and also carry an element of design liability.

The choice of floor solution will also depend upon the time available for construction and the flexibility for later amendment required by the completed project. A timber floor has more built-in flexibility than an *in-situ* concrete solution that would require structural support if new openings were required.

If a pre-cast concrete solution has been selected the units tend to typically bear upon a 150 mm steel angle and a 50 mm structural screed is usually laid to provide an integral floor solution.

10.7.4   2C Roof

Extent: Roof structure (construction), roof coverings (finishings), roof drainage (gutters and downpipes) and roof lights.

Elemental unit quantity: Area on plan of roof measured to eaves.

Sources of error within the cost data

Roofing works, due to their composite nature, are often carried out by a number of subcontractors with a range of resources. A traditional timber trussed roof with felting and tiles would use a specialised supply for design and supply of trusses, resources required to store the trusses and fix in place and then a roofing subcontractor to provide the materials and resources to fix the felt, battens and tiling.

The local markets for materials and labour may have an impact on the availability of resources and a consequent impact on the likely costs. The use of oil-based products for some types of roof, such as bitumen and asphalt, may also make the cost subject to volatile movements in the costs of the raw materials. It may be difficult to capture this volatility through the use of cost indices and the construction economist should ensure a good understanding of the market economics of such specialised materials.

Supply chain procurement

The decision to subcontract specialised works of a roofing nature are dependent upon the technologies required by the design team and the constructor capability.

The availability of contractors to carry out particular roofing solutions may be limited due to product and locational specialisations and this can lead to increased prices as competition may be limited or the need to travel a distance to the project imposes a cost penalty.

The roof in the case study is a curved system. It is to be a proprietary standing seam insulated metal deck. The installation of this system could be either from an approved subcontractor (using a similar procurement method to the steel frame element) or it could be specified, quantified and carried out by a cladding subcontractor contracted as either a domestic or works subcontractor.

Technology

The cost associated with the roof can be related to floor area and perimeter length. The fewer the number of storeys the higher the cost per m^2 of this element as a proportion of gross floor area. The costs associated with this element can be considered under the following headings:

- roof coverings
- perimeter design details, i.e. soffit details
- rainwater goods
- specialist roofing/glazed canopies.

Past quotations for similar systems would be extremely useful. However, cladding systems and associated technologies are continually changing and it is often difficult to gather the necessary information.

10.7.5   2D Stairs

Extent: Stair structure (ramps, stairs and landings), stair finishes and stair balustrades and handrails.

Elemental unit quantity: Number.

Sources of error within the cost data

The errors that may occur are related to the extent of design. Often staircases are designed as the project develops and a prime cost or provisional sum is included in the documentation at tender stage. This allowance is carried into the cost analysis and can be unrepresentative of the true cost. For staircases that are designed and priced within a tender the costs may be difficult to assess due to the often unique nature of staircase solutions.

Supply chain procurement

The resources employed to complete this element of the works will depend upon the technology. If a steel staircase solution was to be used, a steelwork subcontractor would have the resources to design, fabricate and install the staircase. If an *in-situ* concrete staircase was required the constructor would provide the labour to complete the formwork and pour the concrete and a specialist may be used to bend the reinforcement.

Technology

The proportional costs of staircases to total costs are often fairly minor, as examination of the previous analyses has shown. In industrial buildings, such as the case study, the staircases tend to be either steel or concrete either pre-cast or cast *in situ*. The costs associated with them include the structure, any landings required, finishings, balustrades and hand railing.

Reader reflection

Search through your price books to contrast the differing costs of timber, steel and concrete stair solutions.
Think about the fire protection required for timber staircases that is integral in a concrete solution.

10.7.6   *2E External walls*

Extent: External enclosing walls including those to basements but excluding items included with 'roof structure', chimneys, curtain walling, vertical tanking, insulation and applied external finishes.

Elemental unit quantity: Area of external walls excluding window and door openings.

Sources of error within the cost data

External wall costs can be very dependent upon the costs of labour and the conditions in which the work is carried out. As the work involves the completion of the external envelope of the building and the work can be very labour intensive the costs can vary significantly with the weather conditions. The constructor would allow an element within the programme to allow for time lost, reduced production outputs or protection required if the work was to be carried out during the winter months. The allowances are impossible to derive from the historic cost analysis data and little research has been carried out to ascertain the causal impact weather has upon external wall costs. The construction economist should be aware of when this element of the works may take place and consider making an allowance in the cost target.

Supply chain procurement

The external wall costs are a significant element in the total costs that requires careful consideration. The costs of labour-intensive methods, such as brickwork, are rising at significant rates due to the boom in house building in the UK at present. When combined with the reduction in the number of craftsmen available, it has led to premium rates being paid. The enabling works required to lift the men and materials to a suitable height in order to install the works can also be a significant cost which is both time and method related. For cladding solutions, a gravelled and stable area is often required around the perimeter of the building in order to allow access for mobile hoists and platforms. The construction economist should be aware of any limitations regarding access as this could lead to additional costs being included in the tender.

Technology

There are a wide range of external wall construction solutions available ranging from load-bearing brick and blockwork to steel cladding solutions. The final solution will depend on a number of factors not least of which will be the aesthetics of the completed building. The external envelope construction forms the public 'face' or aesthetics of the building.

The case study design utilises a large extent of high technology anti-sun glazing in powder coated aluminium frames. This could be a significant cost and one that should be researched carefully. It is relatively simple to apply rates for brick and block wall construction. In the case study, however, this type of construction forms a relatively low proportion of this element cost. The *pop ratio* is an expression of the efficiency of external wall enclosure and it can be calculated for the project. This is shown on the website.

10.7.7   2F Windows and external doors

Extent: Windows (sashes, frames, linings and trims, ironmongery and glazing, shop fronts, lintels, dpcs and reveals) and external doors (doors, fanlights, frames, linings, lintels, dpcs and reveals).

Elemental unit quantity: Area of openings in walls or enumeration.

Sources of error within the cost data

The costs of this element tend to be essentially component related and as such are very dependent upon the quality of the component specified. The construction economist must develop a sense of the quality that is required by the design team and review previous analyses to establish typical cost/quality ranges. The difficulty this approach embodies is that design teams often specify particular products on the basis of past experience. This specification may be for local manufacturers or for special/non-standard components and consequently making inferences for cost targets may be difficult. The cost of fixing the components is often relatively minor.

The construction economist must determine in a curtain wall system whether windows and external doors are measured and costed separately or included with 'external walls'.

Supply chain procurement

These tend to be specified by the design team and installed by domestic subcontractors. The costs are generally related to the component costs with the costs of installation relatively low. The cost of these depend on the performance and quality specified and the extent of openings in the external wall envelope. If doors are to be automatically controlled and have specialised services they may be supplied and fixed by a nominated or selected contractor.

Technology

Door and window technologies rely on a variety of materials for the fabrication of the frames. Aluminium, steel and plastic are all used extensively in the industry. The most expensive and durable products tend to be powder-coated aluminium.

The cost of fastenings and types of glass may also be significant and relate to the security and function of the building.

Case study

The external windows are taken into account with the external walls element as they form a significant part of the cladding system. The main cost check is the provision of a high quality glazed revolving door and the loading bay door to the production area.

10.7.8   2G Internal walls and partitions

Extent: Internal walls, partitions and insulation, chimneys, screens, movable space-dividing partitions and internal balustrades.

Elemental unit quantity: Area of internal walls excluding openings.

Sources of error within the cost data

The cost data contained in the historic analyses for this element are less susceptible to error as the technology is usually relatively simple and the costs of production are unlikely to be affected by weather conditions. The extent of internal walls and the area for their construction may have an effect upon the cost included by contractors in their tenders. The construction economist may be able to gain an impression of the likely extent of internal walls and partitions through review of elemental unit quantities and rates from historic analyses. The density of partitioning in internal space is more difficult to establish. However, the publication of project drawings by the BCIS may help the construction economist in arriving at a judgement of the likely area in the project proposal.

Supply chain procurement

Blockwork and *in-situ* concrete tend to be carried out by labour-only subcontracts with the materials and plant being supplied by the main contractor. The costs are consequently sensitive to the supply of skilled labour in a similar fashion to external brickwork. The costs of transportation of resources to the various areas within the building remain hidden from the construction economist. The cost of block work to a ground floor partition would be significantly lower than a similar partition in a stairwell on the eighth-floor of an office block. Average costs per square metre in multi-storey work may have to be treated with caution and need adjustment for the project in hand. If the work is to be of a specialised nature it is likely to be subcontracted. Difficulties arise when trying to find comparable elements from historic projects as the specifications vary considerably between the various proprietary design solutions.

Technology

Stud partitioning can be either made from softwood and skimmed plasterboard or a specialist contractor may install a proprietary system. The costs vary significantly depending upon the performance required of the partition. The internal walls can be either load bearing or non-load bearing and the consequent design will be influenced by this fundamental factor. Other factors that would be considered at the design stage could be:

- fire retardant
- aesthetics
- sound and heat installation
- security
- dead weight of constituted wall/partition
- use of space.

The design options available to the designer of the walls if they are to be load bearing are:

- blockwork
- brickwork
- *in-situ* concrete.

Case study

Have a look at the intended use of the space and consider the types of partitions that are required to divide the various areas.
Compare the rates in your price book and calculate the relative differences for each design solution.

10.7.9   2H Internal doors

Extent: Doors, fanlights and sidelights, sliding and folding doors, hatches, frames, linings and trims, ironmongery and glazing, lintels, thresholds and reveals.
Elemental unit quantity: Area of openings or enumeration.

Sources of error within the cost data

The quality and quantity of the doors in previous analyses should be considered when arriving at the cost target. This can only be derived from a review of the specification information and the elemental unit quantities; these may not be available on all the analyses found from the search.

Supply chain procurement

Internal doors are generally supplied by a specialist supplier and fitted with labour-only or contractor's labour. The costs of this element are similar to doors and windows and related primarily to the component costs and, therefore, quality is the important cost driver.

Technology

The costs of internal doors is very much dependent on the quality specified and their use. A hardwood faced internal door that has fire resisting quality would be considerably more expensive than a single hollow core plywood framed door.

The criteria that are taken into account by the design team when specifying doors are as follows:

- functionality, i.e. fire resistance
- aesthetics: related to the use of the internal space
- usage: high volume traffic
- security: extent of door function
- opacity, i.e. allowance of light.

Case study

Review the drawings. Can an elemental unit quantity be derived for this element? Calculate the elemental rate and compare with historic analyses.

10.7.10   3A Wall finishes

Extent: Preparatory work and finishes to surfaces of walls internally.
Elemental unit quantity: Area of finishes.

Sources of error within the cost data

The difficulties in using historic analyses in developing cost targets for this element are similar to that of internal walls. It is often difficult to derive elemental unit quantities from historic analyses.

Supply chain procurement

Domestic subcontractors are invariably used for this work, unless it is of a specialised or proprietary nature.

Technology

The proportion of cost of the internal wall finishes to the overall cost plan is obviously determined by the extent of internal wall division. This is related to the 'density' of usage of the enclosed building and was discussed earlier in Chapter 5. The specification for wall finishes tends to be design led, i.e. the aesthetics of the finished product. However, other criteria such as costs of maintenance, fire resistance and water resistance should also be considered. Probably the most used wall finish is 13 mm thick, two-coat plaster to blockwork. However, the relatively low cost of plasterboard is currently making this an increasingly popular alternative to the more labour-intensive *in situ* work. Large plasterboard sheets are glued to the blockwork wall, the joints taped and scrimmed and in some cases the extent of preparation required prior to decoration is minimal. In most cases a final skim coat is required. Dry wall linings used in the USA and other countries may be increasingly adopted over the next few years.

10.7.11   3B Floor finishes

Extent: Preparatory work, screeds, skirtings and finishes to floor surfaces.
Elemental unit quantity: Area of finishes.

Supply chain procurement

Domestic subcontractors are invariably used for this work, unless it is of a specialised or proprietary nature.

Technology

The floor finishes can also be an extensive cost element that is very much influenced by the intended use of the building. Similar criteria can be applied to floor finishes as those considered in the preceding section for walls. Wet areas such as kitchen and toilet areas would tend to have either a vinyl or quarry tile base to a screed with a waterproof skirting. The floor finish to the other areas is similarly dictated by the functionality of the room's use.

10.7.12   3C Ceiling finishes

Extent: Preparatory work and finishes to surfaces of soffits. Construction and finishes of suspended ceilings.

Elemental unit quantity measure: Area of finishes.

Supply chain procurement

Domestic subcontractors are invariably used for this work, unless it is of a specialised or proprietary nature.

Technology

The cost of ceiling finishes are relatively easy to quantify. The elements can be either suspended on hangers or a frame or fixed directly to the structural background, i.e. plasterboard to the timber floor joists of domestic construction. Similar design criteria of aesthetics, fire resistance (spread of flame) acoustics, maintenance and access for services would all be considered in the design of this element.

10.7.13   4A Fittings and furnishings

Extent: Fixed and loose fittings and furniture including shelving, cupboards, wardrobes, benches, seating, counters, ..., blinds, blind boxes, curtain tracks and pelmets, blackboards, pin-up boards, notice boards, signs, lettering, mirrors, ..., ironmongery, soft furnishings, works of art, equipment (non-mechanical).

Elemental unit quantity: Often these are not measured. May be enumerated or lump sum allowance(s).

The client is also provided with a choice in the early design stage that will involve whether a project is to be 'fitted out' or simply to be a shell-and-core. In most office blocks the end user will 'fit out' the office with a layout that suits their purposes. In this case the essential components that would be allowed would be toilet cubicles and kitchen fittings. In spaces such as these, the costs of fixtures and furnishings may be considerable. The building user may provide all other fittings that could range from being as extensive as internal partitions to minor (and major) items of office furniture.

It is important to clearly distinguish which furniture and furnishings are provided as part of the main building contract and those provided by the client (or others) as a separate contract, usually after the building works are complete.

10.7.14   5 Services

Sources of error within the cost data

The costs of the services elements of a building is increasingly becoming a larger component due to the increasing amount of technology required to support the users' business, leisure and domestic functions.

The design of the other elements will have taken account of the service requirements. The frame for example would have an allowance for service dead load such as air conditioning and the weight of plant items; the floors included in the case study have been designed to allow for access for cabling and trunking.

This element is the most difficult to arrive at a cost target for as discussed earlier. The elemental costs contained within costs analyses tend to be based upon lump sums given by specialised contractors. These lump sums are often not broken down to meaningful elements and few construction economists have the knowledge to assess the accuracy of the estimates. The performance of the services is a key issue and the construction economist must try to establish the costs of services to provide similar performance in buildings of a similar function. The heating costs, for example, may be low pressure hot water and the performance of this system would tend to relate to the extent of gross floor area, the usage of the building and the performance of the external fabric. Even for this simple example it is difficult to match two production facilities such as the one in our case study. For instance, the levels of heating in the production facility would be less than that required in the office.

Log on to the website to view details of the services and consider the service requirements.

The BCIS standard form of costs analysis breaks down the element into the following headings:

- 5A Sanitary appliances
- 5B Services equipment
- 5C Disposal installations
- 5D Water installations
- 5E Heat source
- 5F Space heating and air treatment
- 5G Ventilating systems
- 5H Electrical installations
- 5I Gas installations
- 5J Lift and conveyor installations
- 5K Protective installations
- 5L Communications installations
- 5M Special installations
- 5N Builder's work in connection with services
- 5O Builder's profit and attendance on services.

These services will now be considered and in many cases grouped together for this early stage in the cost planning process.

5A Sanitary appliances, 5B Services equipment, 5C Disposal installations and 5D Water installations

Elemental unit quantity: Number of appliances.

An allowance would be made for provision of toilets, wash hand basins, urinals, and any electric showers that are required and associated pipe work for hot and cold water and for disposing of waste.

The main difficulty with cost planning and this item is that it tends to be designed in performance terms with an outline brief given to the tenderer. Often a lump sum with minimal supporting cost information is incorporated into a price for a contract. Consequently, the detailed price information remains hidden down the supply chain.

5E Heat source, 5F Space heating and air treatment and 5G Ventilating systems

Elemental unit quantity: Due to difficulty of measurement they are often related to the floor area they are servicing.

These elements also tend to be designed by either a mechanical and electrical consultant or a specialised contractor. They are often specified in performance terms that can be difficult to cost plan and check. An allowance is usually made on the basis of a cost per m^2 and the costs qualitatively adjusted for the performance of the facility and individual space and building requirements.

Reader reflection

The reader is recommended to refer to price book information and also browse the specification information on the cost analyses to get a feel for the types of service installation and the costs per m^2 of GFA.

Assess the percentage service costs form of the total costs of various types of buildings.

5H Electrical installations

Extent: Includes electric source and mains, electric power supplies, electric lighting and electric lighting fittings.

Elemental unit quantity: Number of electrical outlets.

The costs of this subelement are dictated by the extent of the power supply required by the building and its distribution within the building and also the extent of power and lighting required in each space.

A cost per m^2 is generally taken from the previous project and a qualitative adjustment made to take account of the increased/decreased number or density of outlets.

5I Gas installations

The costs of this subelement are dictated by the extent of the power supply required by the building and its distribution within the building and also the extent of power required in each space.

A cost per m^2 is generally taken from the previous project and a qualitative adjustment made to take account of the increased/decreased number or density of outlets.

5J Lift and conveyor installations

Extent: Includes lifts and hoists, escalators and conveyors.

Elemental unit quantity: Number of lifts, escalators and conveyors.

A specialised manufacturer who will supply and install the lift to a prepared lift shaft with a specified power supply carries out the installation of a lift. The costs associated with the lift are the structure, electricity supply and ancillary requirements, such as a plant room, etc.

5K Protective installations

Extent: Sprinkler and fire installations and lightning protection.

Elemental unit quantity: Gross floor area.

These tend to encompass fire protection and lightning protection and can vary greatly in cost depending on the use of the building and the regulations that pertain to its use. If the building is to contain hazardous inflammable material, for example, it may need an extensive sprinkler system.

5L Communications installations and 5M Special installations

Warning, visual, audio and special installations may have to be considered carefully in some building types and particular locations.

5N Builder's work in connection with services and 5O Builder's profit and attendance on services

These allowances on mechanical and electrical services are often expressed as a percentage and shown separately in these elements wherever possible.

10.7.15   6 External works

The external works are included in the BCIS analyses as the total costs but are not related to gross floor area or building quantity. They can be considered under four sections:

(1) Site works
(2) Drainage
(3) External services
(4) Minor building works.

6A Site works

Elemental unit quantity: Often not measured, but expressed as a lump sum or percentage of total cost.

This heading includes site preparation, surface treatment, access roads, car park areas, planning, site enclosure and division and fittings and furniture (signage and seats).

6B Drainage

Elemental unit quantity: Often not measured, but expressed as a lump sum or percentage of total cost.

The cost of drainage is related to the site location, building location on the site and extent of external services. The extent of enabling works that may need to be carried out to existing sewers should be investigated, as they may not have the capacity for increased loading.

Any car parking would need a system for rainwater collection and disposal.

6C External services

Elemental unit quantity: Often not measured, but expressed as a lump sum or percentage of total cost.

Water, fire, gas, electricity and telecommunications have to be provided to the building. The allowances made in the cost plan would be based on the cost planner's knowledge of the location of these services in relation to the site. If new substations are required to be constructed or gas pipelines extended over a long length of ground, this could contribute to the high costs of this element. Builder's work for these items should, wherever possible, be shown separately.

6D Minor building works

Elemental unit quantity: Often not measured, but expressed as a lump sum or percentage of total cost.

The allowances are for providing minor works in relation to the external services.

10.7.16   7 Preliminaries

Elemental unit quantity: Often not measured, but expressed as a lump sum or percentage of total cost.

Sources of error within the cost data

Access to the resource cost information from historic projects tends to remain in the contractor's domain. The contractor's estimators would price the resources required to programme and integrate the subcontractors, provide supervision, site facilities and insurances based on an agreed programme for the project. The costs can be considered as method related and can be broken down into *fixed costs* such as enabling works, *variable costs* such as electricity usage and *time related costs* such as supervision of the works. The majority of preliminaries costs tend to fall into this latter category. The preliminaries element of a contractor's tender can be considered as the main area of variability between competing contractors. Consequently, the detailed build-ups incorporate the contractor's key competitive knowledge of how the project may be constructed, innovative forms of temporary works and methods of shortening the programme and remain confidential.

Often prices of individual preliminaries items within the bills of quantities or contract documentation form the basis for evaluation of claims for delay and disruption. This influences the tender strategy and can lead to rate inclusion that do not represent the true costs of the resources. The construction economist has to be aware of the typical costs of the major items of preliminaries, such as staff cost, cabin hirage and typical levels of cost of insurance. This knowledge is derived through experience of pricing levels on previous projects and reference to published materials in technical journals and price books.

Supply chain procurement

The major costs of subcontractors involved in preliminaries works tend to be either enabling works such as temporary roads or access or temporary works such as scaffolding or formwork provision. When pricing these items the contractor takes a commercial decision that is influenced by cash flow, possibility of future returns and risk upon where to include the monies within the tender. The costs of inclusion of a project *intranet*, for instance, would be allowed for here. It is unlikely however that a project of the size of the one within the case study would have such a facility. The lesser cost items that subcontractors need (or must provide themselves) are the provision of services such as electricity, telephones and gas supplies.

The development of a cost target for the preliminaries section of the works requires a good understanding of the cost drivers of the contractor's method, related charges and how a particular project influences such costs. The construction economist reviews previous project preliminaries costs. As the major cost bearer is time related, project analyses of a similar construction period and similar technologies would be reviewed to gain an impression of the broad range of costs for this element. It is important to consider how project characteristics may have influenced the preliminaries costs. For example, a project that was adjacent to an existing building that required shoring up may be excluded from consideration in the case study as these preliminaries costs would not be incurred. The construction economist compiling the historic analysis recognises this weakness of the information and will often state the amount included for such an item in the project details. However, as mentioned above, it is difficult to assess the accuracy of the allowance with the tender price.

Log on to the website to view details of how the authors have built up the preliminaries cost target.

10.7.17   *Employer's contingencies*

The employer may wish to allocate some funds to an area of the cost plan which reflect unresolved areas of the design brief and can help administer the project post contract. The scale of the contingency will reflect the complexity of the project and the maturity of the design brief. This element has no definition within the SFCA, however, and needs to be carefully defined by the design team and client to ensure that its scope is understood by all parties.

10.7.18   *Design fees*

This element is allowed for projects procured using a design and build form of procurement. It is divided into two subelements:

(1) *Work that is completed before the commencement of construction.* This would involve developing the client's brief to a stage that could be used for the drafting of

the client's requirements. The client's requirements form a tender and contract document under the design and build procurement route. Other duties
would involve assessment of the contractor's proposals. The reader is referred
to the RIBA (2000) plan of work, which identifies clearly the design team's
role under this form of procurement at preconstruction stage.

(2) *Work during construction.* This may involve acting as a design professional working for a contracting organisation. If the designer was novated, the construction
design costs could be assessed. These may include detailed development of
each element and could be included here. The reader should review the
RIBA (2000) plan of work and the detail on the website which discusses how
the advice would change if an alternative procurement route to the sequential
traditional route adopted by the case study was chosen.

10.7.19   *Development of the detailed cost plan*

The detailed cost targets are now used to formulate the detailed cost plan. The
construction economist's judgement is based upon experience and costs from
previous projects' cost analyses. It should be stressed that this process of providing targets is a dynamic and an interactive one and is not a passive reaction to
merely accepting without question or further research the information provided
by the design team. As the cost targets are developed the emerging design will be
value engineered to test the assumptions and effectiveness of the design team's
decisions on the functionality of the emerging design. The construction economist has a large role to play in this process using knowledge of how other design
teams have solved similar functional problems. Drawing from studies of the
details of past projects and informed by their economic consequences, the construction economist can make a large contribution to the design choices made.

The reader is encouraged to critically look again at the drawings, review the
details of the previous cost analyses and then consider the function of each of
the case study elements and relate it to the user's requirements. For example, if
the user's requirements for access to the two floors and production facility are
considered, the construction economist may start to examine the provision for
staircases, their location, and the necessity for a lift. It is by this systematic analysis
of the design that the design team can develop better solutions. The construction
economist should have a good feel for the extent of design risk that has been
allowed for within the cost plan. This is gained by systematically working through
each element and assessing the availability of information from the design
team and the firmness of the client's brief. The allowance for price risk should
always be kept in mind as the design develops. Any slippage in the preconstruction programme may lead to a later tender date and an adjustment to the
allowance made may have to take place. A 3-month change in tender period
date may have a large impact upon the tender price over and above that reflected
by the Tender Price indices. The construction phase may move to a period
of poor weather and an increased allowance may be made by the tendering
contractors.

The next stage of the process is to cost check the elemental targets using a more
detailed method of quantification and cost build-ups. This process of cost checking
will be discussed in Chapter 11.

10.8   Reader reflections

- Review the elemental cost analyses on the website and highlight the features that are similar to the case study. How would you take these differences into account when arriving at a cost target for each of the elements?
- What problems do you consider are inherent in the techniques of applying Tender Price index and Locational index adjustments to reflect Price Risk? What are the local market conditions that apply to your region at the moment?
- When considering Design Risk, how would you analyse historic analyses to ensure that a sound understanding of the quality of the project's designs is developed?
- Review the elemental cost analysis information. What can you deduce from the preliminary costs of these past projects?
- The processes discussed in this chapter have focused on pricing a design. How would they change if the design team were designing to a price?
- The processes of design cost management rely upon a great deal of discussion between the design team, client and construction economist. How would you record these flows of information?

11 Confirming the cost targets: detailed design stage

11.1 Introduction

This chapter aims to:

- Introduce the reader to the cost checking process
- Discuss the quantification required at this stage
- Illustrate how the cost check for the substructure element may be carried out
- Highlight the need to consider the consequence of undesigned elements and the management of contingencies
- Summarise the difficulties of sourcing reliable cost information.

The cost checking process involves checking whether the elemental target allowances in the detailed elemental cost plan are realistic. This involves a simulation of the processes that the contractor undertakes when tendering from a substantially completed design. Approximate quantities are generated from the drawings and all-in unit rates are built up and applied to these quantities. The standard form of cost analysis elemental definitions are used to allocate costs in the appropriate categories. The extensiveness of the design will vary from project to project and the procurement route adopted. One of the reasons for the rise in popularity of the design and build procurement route was that it reduced the time from briefing to start on site as the design process and construction process could overlap. The working drawings for the later sections of the work may be unavailable at the early stages of construction and the elemental cost targets may still be within the cost plan after work has commenced upon site. This often occurs for the later sections of work such as finishes and external works. As indicated in Chapter 5, the amount of unallocated expenditure decreases exponentially with the increase in design information. This can mean that if the construction economist has made errors in the development of the cost plan then the design team will have little flexibility in later decision making. Often savings are sought in the most visible areas in a building, such as floor and wall finishes, when cost targets are exceeded. This is paradoxical as these elements often form a proportionally small expenditure and cost savings will not be significant without a major reduction in quality. This has been cited as being one of the problems associated with quality in the design and build procurement route as the contractor has designed down to a market price and has little flexibility to find savings as the design develops.

11.2   Cost checking

11.2.1   *Design information and quantification*

The construction economist must use a knowledge of technology in order to quantify the cost components for each element being checked. Decisions upon which parts of the design to be quantified are based upon the judgement and the time available to carry out this process. If the project is procured using the traditional method, the construction economist may well be employed in producing contract documentation such as the bills of quantities, specifications and forms of tender. This would leave little time to carry out the cost check. However, it is at this stage that the tender price of the contractor can be predicted with the most accuracy and it also gives the design team the opportunity to carry out any late changes to the design to bring it back to plan if variances are found. The detailed quantities could, however, be used as the basis for the cost check. If the original cost targets were based upon approximate quantities and the details of any design changes have been recorded by the construction economist, the process of cost checking can be speeded up. The allowance included within the elements for design risk should continually be reviewed as the detail of the design develops. If the design risk is no longer present then the allowances made would be redundant and could be used elsewhere.

Construction price books often have all-in rates that can be applied to approximate quantities. It is important for the construction economist to use the most contemporary information available as price book information is often very generalised, out of date, based upon regional supplier and labour markets and unrepresentative in the allocation of overheads and profit. Priced bills of quantities from previous projects may be useful as a basis for the development of all-in rates to be applied to approximate quantities. However, they should be used with caution as they can be unrepresentative of the real cost and can be regarded as an arbitrary allocation of prices to arrive at a market price.

11.2.2   *The cost checking process*

It is important for the construction economist to record the decisions made by the design team as the design develops. This allows an audit process to be undertaken as the project proceeds to highlight the critical decisions which have led to cost commitments. This recording of the contributions that the design team make is important, as it will help engender commitment to the cost management process and reinforce the link between design decisions and cost consequences. The construction economist must be systematic at this stage and record the detailed drawing and specification details that the cost check is based upon, the date the check was carried out, any variance identified from the original cost plan and the overall impact the revised design details have upon the cost of the project.

11.2.3   *Management of the variances between cost targets and cost checks*

Any arrangement for control requires four elements:

(1) a standard by which to compare actual with predicted performance
(2) a means of measuring or detecting the actual performance

(3) a means of communication if a variance from the predicted standard has occurred

(4) a mechanism that allows action to take place to bring the actual back towards the predicted.

Figure 11.1 shows a cost checking routine comprising three pathways aligned to the given cost circumstance of the project.

As discussed earlier, the standards that are used throughout the cost planning process are derived from historic cost information, which is captured through the use of priced market information. For reasons cited earlier, these standards may be poorly founded (or even incorrect for the planned project) and may give rise to issues of variance as the cost check proceeds. In order to ensure that the construction economist learns from the cost prediction and checking processes, a systematic testing of standards and reflection of the actual against the predicted should take place.

The measurement for detecting actual performance is still error prone as the cost information applied at the cost checking stage may contain similar errors to the historic cost analysis information such as unbalancing, loading and market variances. In order to mitigate against these, the construction economist must review cost information from a range of sources prior to being applied to the approximate quantities.

The mechanism of communication of the results of the cost check must also not be overlooked. Design team meetings should be convened with all parties responsible for the management of the design process being present. The reasons for variance from the original cost plan should be considered and also an action plan to bring the project back to plan should be discussed and agreed.

The responsibility for action to respond to the variances highlighted by the cost checking process should be with the individual managing the design process. This may be the construction economist or a project manager or architect. The extent of action that can take place after the variance has been identified will depend upon how much design has been completed, the time available for element design review prior to construction, the commitments already made, the organisation of the design team and the project procurement strategy. This is a complex area to manage and should have been considered in advance of the design commencing. Clear lines of responsibility and authority should have been highlighted by the design manager and the processes to be undertaken in such circumstances clearly noted.

There are many examples where the process of design cost management has gone spectacularly wrong. Unfortunately these are highlighted, particularly by the media, whereas good examples of cost control are not so newsworthy.

Figure 11.2 shows an examplar cost check of the substructure element of the case study that is based upon approximate quantities.

Log on to the website to view the elemental cost checks.

The reader is referred to the cost check quantities for the various elements and is encouraged to apply price book information to carry out his or her own checks to gain a better feel for the accuracy of the elemental cost targets.

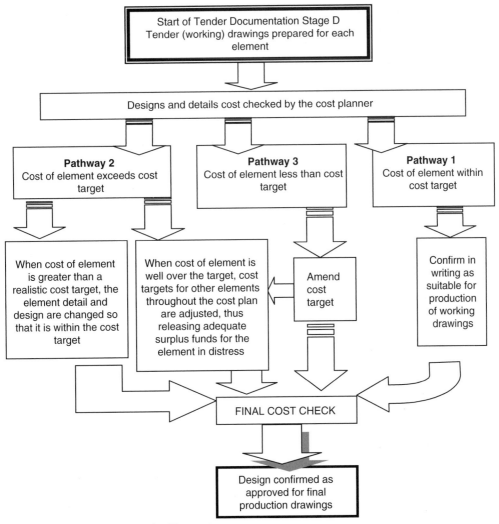

Figure 11.1   Pathways to cost checking.

The cost check carried out in Fig. 11.2 shows a significant variance in projected cost from the original cost plan. A variance of this magnitude could cause the design team major problems. Although the relative percentage is within the expected range of accuracy for the whole project (i.e. 3.14% on total project allowance) it is a relative increase of plus 128% on the original substructure figure! This would require a great deal of further analysis and discussions with the engineer and design team to identify the reasons for such an abnormal overexpenditure.

Reader reflection

Identify the reasons for overspend on the substructure element of the case study.

Job No. 123 Cranium Containers	Client – Cranium Containers Ltd	Cost plan no. 2	Date
Total project cost from cost plan 1	£1 100 000	Costs per m^2	£1038
Element title	Substructure		
Element no.	1A		
Preliminaries apportioned	No		
Original elemental cost			
Item	Description	EUQ	Unit
1A	*In-situ* concrete edge beam, including excavation, disposal, reinforcement, hardcore, earthwork support, backfill 600 × 1000 mm	120	m
B	Column bases. Including excavation, disposal, reinforcement, hardcore, earthwork support, backfill 600 × 1000 × 1000 mm	16	Nr
C	Column bases to columns roof overhang supports as above, 600 × 600 × 600 mm	4	Nr
D	Ground slab, including excavation, disposal earthwork support, reinforcement, visqueen, hardcore, powerfloated	528	m^2
E	Provision of lift pit		item
F	Excavate to remove top soil	550	m^2
G	Removal of mound	150	m^3
H	Mass fill concrete	198	m^3
	Revised element cost		
	Revised project cost plan total		

Figure 11.2 Substructure cost check.

The reader is encouraged to review the other cost plan elements and using the drawings and specification details carry out a number of cost checks. The website can be used to view the cost checks completed by the authors.

The construction economist has to consider the risks of the undesigned elements at the cost planning stage and will either include a contingency allowance as a single item within the plan or spread the contingency across the range of elements. The former approach is the preferred method as it highlights the total amount of unallocated expenditure and can be managed in a more transparent fashion. A large contingency element can lead to problems, however, as it can be seen as a security net by some members of the design team and used to cover problems that have occurred due to taking short cuts from rigorous design procedures and best practice. The management of the contingency allowance should be similar to the other element allowances in the detailed cost plan. The risks associated with overspending/underspending on the various elements can be predicted from past records and the design risk allowance managed in a manner that reflects the proportion of risk outstanding. For example, it would be poor practice to have a fully completed detailed design, to have received the tenders within the cost plan allowance and to have an unspent allowance for design risk as the construction was heading for the completion of the finishes stage.

The management of unallocated expenditure is of particular importance in the privatised utility market in the UK. The utilities are regulated by a government appointed Office of Fair Trading that oversees expenditure on capital repairs and maintenance. This expenditure is budgeted and monitored each financial year. A particular utility company has a sophisticated system of cost prediction, the accuracy of which is regularly monitored and is generally accurate within ±5%. However, the company must keep a contingency allocation to a minimum as it is viewed as unallocated expenditure and consequently does not count as capital expenditure in the eyes of the regulator. As a consequence of this position, the contingency allowances are closely linked to identified risks and as the project's design develops the need for the specific contingencies are reviewed. If the risk is no longer likely to occur, the contingency will be released for expenditure on another project. This aspect of risk analysis of the design process and contingency management is a practice that should be borne in mind throughout the cost advice process.

The reader is referred to Hayes and Perry (1986) and Raftery (1994) for further information.

11.3   Summary and limitations of the design cost management process

As a guide to the construction economist the following factors should be borne in mind in practice:

- There are likely to be problems with obtaining the data. That is, it is rare that an element reflects the product and the process of construction.
- Expect there to be difficulties in being able to directly compare the cost solutions of one element with another on a similar building, due to the interdependency of design solutions and the difficulty in identifying the causal

factors that lead to the cost consequences of such solutions. It is virtually impossible to carry out meaningful comparisons of different technological solutions to design problems. These remain at a relatively simple level.

- Strictly speaking, there is no contractual responsibility for the design team to meet cost targets. However, there are many other pressures to perform to target, particularly if the contractor is a design and build contractor.
- Appreciate there are problems of getting past the market barriers for information as the contractor submits market prices that are unreflective of cost and this is also reflected down the supply chain. The best method to solve this problem would be through contractors developing supply chain clusters that had transparent information systems that could be shared via the web. The contractual and trust barriers to this are slowly coming down and may be prevalent in the future.
- Finally, the cost plan and cost check are still only predictions of market pricing of complex designs with methods that are fairly crude. Unless we research into means of collecting the data in a non-threatening manner to the political and contractual relationships set up by the supply chain we will always have to rely on the relatively unsophisticated methods currently available.

11. 4   Reader reflections

- What are the difficulties associated with cost checking when one considers specification information?
- How would you communicate the results of the cost check if there was a significant variance from the cost targets?
- What action can be taken under the various procurement routes if a project looks like it may well be exceeding its budget?
- How would the extent of design development and commitment to project constructors limit the extent of such action?

12 Design cost management and the future

Introduction

This chapter aims to:

- Review the current practice of design cost management
- Discuss how classification theory may make a contribution in the future
- Suggest how information technology developments may affect the practice of design and cost management
- Consider how e-business and procurement changes in the construction environment may drive change
- Suggest an approach that may start to separate price from cost data.

This book has described how design cost management is practised today. It has traced its development and the vital contribution that cost modelling makes to enable the construction economist to play a significant role in seeking to achieve value for money for our clients and the community. We have also demonstrated the need for cost data in a form that is reliable, plentiful and useful. The role of the BCIS has been described as being one of the major custodians of building price enabling it to be accessed and manipulated to allow construction projects to be cost modelled at various stages in the development of the project's design. Through the use of the website the reader has been able to participate in the dynamics of this process by being able to observe design cost management as well as being able to carry out the process for themselves. The authors believe that through such an interactive process the inquiring construction economist can develop a much clearer understanding of the complexities of the process. This understanding embraces the need for a well integrated and committed design team with a clear understanding of the implications of design upon construction, an appreciation of the general economic conditions as they impact on a developed economy together with a detailed understanding of the economic conditions as they impact on the construction industry (Ashworth 1988, Morton and Jaggar 1995).

The process of design cost management as described in this text has been set within the well established procurement method of contractor selection based on lump sum competitive tendering as discussed in Chapter 3. Additionally the RIBA plan of work has been used to relate the timescale and context of the process of design cost management in order to show what, when, how and why it should be done. The purpose of this final chapter is to speculate as to the role and nature that design cost management is likely to take in the future especially as a result of the trends towards more open and harmonious procurement strategies as discussed in Chapter 4. We have also stressed, at various times, the problems and limitations of the source cost planning data, derived from bills of quantities, which are used to populate the various cost models that are used in design cost management.

So what is the future? Let us start with a few facts and trends. As discussed in Chapter 3, the use of lump sum tendering cannot be described as entirely successful as many contracts are delivered late and over budget and there is evidence of disputes occurring leading to long and acrimonious delays in their ultimate settlement. In fact there are various quantity surveying organisations who have specialised in solving disputes in construction contracts such as J R Knowles and Partners as well as a number of text books about dispute resolution together with a CIB Working Commission W087 Post-Construction Liability, concentrating on such dysfunctional contract arrangements.

As a result of these failures there has been, and continues to be, much pressure both within and outside the construction industry to bring about major improvements in the procurement of construction work in order to achieve buildings for the right price of the right quality and to be delivered within the required time frame (Latham 1994, Egan 1998). Chapter 4 has traced these various pressures and changes and has identified the preferred future directions: primarily the use of design and build strategies within a partnering arrangement in order to create a much more open and project orientated approach based on an agreed set of objectives. Its aims are focused on client satisfaction and a shared approach to the resolution of problems rather than the current antagonistic philosophy.

As has been discussed in this book, current design cost control management applications are based largely on *black box* cost modelling techniques where the design team and specifically the construction economist are seeking to best guess the price the constructor requires for the project. One of the major limitations of such opaque techniques is that the various processes going on within such models cannot be observed and therefore any limitations, problems and errors occurring within the dynamics of their operation remain largely unobserved by the construction economist and therefore, the outcomes may often be erroneous, unreliable or misleading.

Perhaps the black box approach idea needs a little more clarification. Traditional black box models are those based on statistical approaches, which we have termed deductive models in Chapter 5. When we operationalise such models we rely on evidence gathered from a sample of data from which, after statistically process-ing, we may be able to draw certain inferences from the evidence so presented. Of course, the major difficulty of such an approach, is that the inferences are deduced and not established from a causal or algebraic relationship and therefore the results may not always be correct. A further limitation is that such models need to be populated by a large data sample which, as discussed earlier, may not always be possible. So why is our design cost management black box still used, particularly when it was suggested that inductive models based on causal relationships were used to establish our cost plans and to carry out our cost checks? The answer is to do with transparency, or the lack of. We are attempting to apply models based on causality, but in essence our basic data unit rates from bills of quantities are essentially black box in that we can only best guess what data have actually made up the unit rates. The problems of using such data have already been highlighted in earlier chapters. Mention has been made of the fact that the constructor's estimators are often working with limited time and little feedback being provided from the planning department of the construction organisation. Standard texts on estimating serve to reinforce the point in that they explain and demonstrate the application of estimating techniques with limited reference to the construction process (Brook 1993, Smith 1998b).

This black box problem is further exacerbated by the fact that the cost estimate established by the constructor is converted into a price but governed by social opportunity costing rather than being directly related to the cost of the resources as discussed in Chapter 1 (Fines 1974). If this is correct then one can see why design cost management can be further distorted by the manipulation of the unit rates by the constructor to aid cash flow or take advantage of inaccuracies in the quantities contained in the bill of quantities. Again, these points have already been highlighted earlier in the book.

12.2   Possible future directions

So how do we move forward to improve the situation given the changing culture towards greater openness and partnership?

It is the view of the authors that the following fundamental changes are necessary if a sustainable spirit of cooperation and better value for money in our building projects is to be achieved:

(1) Resource driven data should form the basis of our cost models in our design cost management.
(2) Such data have to form part of an effective information system as identified in Fig. 4.1 in order that the various contributors to the design and construction process can aim to optimise and implement their particular contribution by being able to manipulate information relevant to their needs without the need to rely on an erroneous translation as currently takes place.

How can these improvements be put in place and how can design cost management take advantage of these changes?

It is perhaps worth stepping back a little to remind ourselves of how history has led us to our current stage of development. When design cost management was first developed over 50 years ago we had to rely on manual endeavours without the aid of information technology. As a result the various construction related professions created a set of manually driven conventions and procedures in order to aid their particular contribution to the design and construction process. We have already mentioned some of these conventions and procedures such as the RIBA plan of work, Standard Methods of Measurement (SMM), standard forms of contract together with the plethora of text books explaining such conventions, both formal and informal, and the procedures that are required to make them work.

Unfortunately, perhaps because of inertia and the resistance to change by the professions there have been limited improvements in formal communications within the construction industry. As a result the power of information technology has not been applied in a holistic and integrated way, but has focused on improving the speed and accuracy of producing information that would normally be produced by the particular professional concerned. Evidence of such limited applications can be found in many products promoting the application of information technology within the construction industry.

We are guilty of looking down our telescope the wrong way. What we need to do is to turn the telescope round and see the role of information technology as a means of providing the infrastructure within which our data reside so that they can be accessed, manipulated and transposed by all the client, design and construction parties into the necessary information requirements to facilitate optimisation and implementation.

Tackling the first problem the solution is actually simple; construction economists should produce cost models rather than price models, or perhaps more fundamentally produce resource models. These are the basis from which the cost models are established. The reason resource models should be the basis of our cost models is because they are free of distortion in the sense that they are the basic building blocks forming the completed structure: bricks, mortar, tiles, etc. (the commodities). The other resources are those required for their location into the structure: labour and plant. Of course, this process does not happen without considerable management expertise especially in complex difficult projects. So, the resource of management is also required as part of our resource model. A further important resource needed, therefore, is that of time within which to carry out the process of completing the structure.

The basic resource model is shown in Fig. 12.1.

The word commodities has been used to reflect the type of such resources that make up this resource set which are:

■ formless materials: cement, aggregates
■ formed materials: bricks, timber
■ components: windows, roof trusses.

The term energy has been used in our simple expression to describe the process of locating commodities because the rate of carrying out work involves distance,

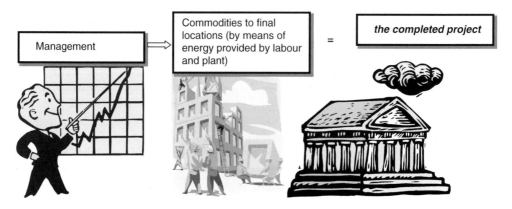

Figure 12.1

mass and time. Clearly, constructing a building is an energy-consuming process, which is primarily concerned with locating our commodities:

- mass (what has to be lifted or moved: the commodities)
- distance (how far they have to be lifted or moved)
- time (the time required to lift or move them).

The above analysis, at first glance, seems straightforward and consistent, but unfortunately it is not. The reason is that, depending on where the supplier/purchaser interface is in the supply chain, the definition of our components will vary and of course the definition of the completed structure will vary. For example, to the manufacturer of roof trusses the completed structure is the roof truss and the commodities are the formed material of timber, the components in the form of nuts and bolts and perhaps any glues or adhesives which are formless materials. To the manufacturer of the nuts and bolts the model can be restated to reflect their particular interests as being only concerned with manufacturing nuts and bolts. However, the basic equation remains valid, but the context varies depending on the supplier/purchaser interface.

The above discussion also explains why the cost/price model varies depending on the context as dictated by its position in the supplier/purchaser interface. For instance, the supplier of the nuts and bolts charges a price for their provision whereas to the purchaser of the nuts and bolts they reflect a cost as they become absorbed in the construction project. When the project is complete it becomes the price the constructor requires which becomes the cost to the purchaser: the client. One can now begin to see why the conclusions drawn by Fines (1974), as explained in Chapter 1, can have some credibility due to the influence of the market on what ultimately the purchaser of the goods (in our case a completed building) may have to pay. One only has to consider the housing market where the relationship between the cost and price is virtually non-existent. What influences the price the owner can charge for the sale of a property is dictated by conditions in the property market. We all know that the price to purchase a flat in central London compared with one, say, in a northern town will be very different indeed.

So, is it worth trying within our information systems to attempt to model resources? The answer must be yes, if for no other reason than the fact that the constructor, at what ever position in the supply chain must produce such a model in order to be able to plan and control the process of construction. In fact, this is precisely what all constructors attempt to do with varying levels of sophistication, ranging from simple bar charts to complex computer driven resource optimisation models based on critical path methods.

We have already discussed the limitations of our current SMM-based price models but we can now explain very clearly from our consideration of energy where distance, or location in our case, and time are fundamental requirements in working out the energy needs in terms of labour and plant—both requirements which are not easily or reliably ascertained from the bill of quantities. We can also now see what the constructor needs to identify in terms of *buildability*; the identification of the least energy commitment by selection of the most appropriate combination of labour and plant. Such a combination is not necessarily that which commits the least amount of plant and labour but that which can operate at an optimum within a given space and can be fully utilised over time. An obvious example is where one plasterer is working in a large space or 10 plasterers are working in a small space; this will not provide an optimal solution. Similarly, how deployable over time a particular resource is on the project will also have a considerable influence on the buildability. For example, a tower crane is an expensive item of plant with high installation and removal costs whether or not it is operating. However, non-specialist labour is clearly much more flexible in its deployment being able to carry out a variety of non-specialist tasks and being easily re-located on other projects. We can now see how influential the design solution is in terms of minimising the energy needs for the construction of the project. Of course this analysis is very much an oversimplification in that many other factors influence the buildability of the project such as how complex the construction process is, how much of the work can only be carried out sequentially, resource availability, etc. Many of these issues were mentioned in Chapters 2 and 5 when discussing the current limitations of the bill of quantities, but our analysis above clearly shows why: the energy requirements in terms of labour and plant are difficult to determine as they are dependent on knowing the distance, or location in our case, of the commodities in the project and the subsequent time needed for their location in the completed structure.

In Chapter 5 we discussed how the rules of the SMM had been changed in an attempt to mirror more closely the process of construction by introducing method-related charges and also by the need to provide certain drawn information and attempting to assist the constructor to determine the construction programme. The answer to the problem is very straightforward, namely, that the resource and financial management of the project should be based on the constructor's programme, which of course is a true resource and cost model of the construction project.

Fact file

Why do we not use resource and cost models based on the constructor's programme? It is impossible to answer the question definitively, but one can speculate as to a number of possible reasons which are:

- The success of the bill of quantities within lump sum tendering as a device for the selection of the contractor based on financial competition.
- The fragmented and litigious nature of the construction industry, which tends to encourage and support opacity rather than transparency.
- The strength and dominance of the design professionals, especially quantity surveyors who do not wish to see an erosion of their influence and power.
- The failure of information technology to take a system's view of information management. Information technology practitioners seem to be more concerned with speeding up and making more accurate current procedures, such as the production and pricing of bills of quantities.

12.3   Some earlier attempts

As mentioned in Chapter 2 a number of attempts have been made to either model the constructor's programme or to allow the constructor's programme to become the basis of resource and financial management. Many of these approaches are variants of the SMM-based philosophy and therefore offer no new or fundamental approach to achieving resource-based cost models. However, two approaches are worth mentioning in that they were radical departures from those based on the SMM and both in their different ways attempted to produce a true resource and cost model. These are:

(1) Operational Bills
(2) British Property Federation system.

However, in attempting to produce such models it is worth stressing the fact that resource identification, in deterministic terms, by anyone not responsible for their provision may not be possible. Commodity resources in terms of how much is required in the finished project can be precisely determined from the architect's drawings. However, the waste on such resources can only be guessed at, the constructor being the only one who can include with confidence such waste allowances based on records, which are under constant review deterministically. When one considers the labour, plant and management requirements, then the construction economist (unless working for the constructor) can only estimate such resources and state the quality of workmanship expected. For example, concrete will be produced by mechanical mixer and certain conditions regarding its place-ment specified, or that a competent foreman will be required to be on site at all times, etc. This partly explains why designer-based construction economists find buildability such a difficult area to consider objectively and explains why design and build procurement approaches are seen to be so beneficial when the implica-tions of design on the construction solution can be more rigorously considered.

12.3.1   *Operational bills*

The first approach was developed in the 1960s under the direction of the Building Research Station (now the Building Research Establishment). The research being undertaken by two of its senior researchers (Forbes and Scoyles 1963). They produced an 'Operational Bill' which attempted to provide a bill of quantities that reflected the various activities or operations[1] required to construct a building together with the sequence within which they should be carried out. Although there are other definitions the spirit of this definition is particularly useful as it conveys the notion of the carrying out of a process which is task driven rather than product driven. The bill was structured to mirror a precedence diagram which indicated the various activities or operations to be carried out, together with their relationship to each other, thereby indicating which were sequential in nature and those which could be carried out in parallel. A further important feature of the Operational Bill was the fact that within each activity or operation the commodities were quantified net in their normal purchasing units, such as rolls of damp proof courses and num-bers of bricks, and labour was presented as an omnibus item describing the amount of work needed to be carried out in order to complete the activity or operation.

---

[1] An activity or operation can be defined as a piece of construction work which can be carried out by a gang of operatives without interruption from another gang (Forbes and Scoyles 1964).

The management and plant resources were presented at the end of the bill to ensure that the tenderer included these items. The reason for their inclusion at the end was that the tendering contractor would have a better view and awareness of the work needed to be carried out in the project and its management and plant implications. In fact the presentation of management and plant was little different than the method related charges currently contained in the current SMM used in both the building and civil engineering industries as discussed in Chapter 5. So why did this development fail? In many ways it seemed to answer all the criticisms identified in this book. A number of reasons can be advanced, but most important were the following:

■ The expertise concerned with the management of the construction process was and still is very much with the constructor and therefore the programme of activities including their sequence did not necessarily accord with the design team's view of the process. This is another way of stating the fact that the design team was not very familiar with the buildability issues as they affect the construction process.
■ There was inertia from the design and construction team, especially the quantity surveyor, who were reluctant to move away from the production of bills of quantities based on the SMM.
■ The notion that the design had to be complete in order to determine the activities and their sequence. As we have already pointed out with existing SMM-based approaches they have the benefit of opacity and as a result the bill of quantities purports to represent reality which is often not based in fact. In Chapter 5 the discussion on defined and undefined provisional sums and the need for certain drawings to be provided with the bill of quantities were developed to remove some of this opacity.
■ Poor timing, in that if the approach were tried today it would have the benefit of being underpinned by information technology to allow rapid manipulations. Revised networks could easily be created to accord with the constructor and to rapidly reflect inevitable changes as work proceeded. This would be a much more useful and dynamic resource and cost model of the building under consideration.
■ Poor marketing, which alienated rather than captured the imagination of the industry at large. Sadly, it received a rather hostile reception.

12.3.2   British Property Federation system

The next attempt to produce resource and cost-based models was that proposed by Barnes under the British Property Federation initiative (British Property Federation 1983). It was a very radical approach which was designed to achieve the following:

■ Change attitudes
■ Produce buildings more quickly and at lower cost
■ Alter the way in which members of the professions and the contractors deal with each other so that each is given a strong incentive to manage his/her contributions with energy, foresight and to work closely with others in the client's interests

- Remove as much overlap as possible between designers, quantity surveyors and contractors (which is prevalent under the existing approach)
- Redefine the risks so that commercial success of the designer and the contractor depends more on their abilities and performance
- Re-establish awareness of real costs by all members of the construction team
- Eliminate practices which absorb unnecessary effort and time and obstruct progress to completion.

All of these recommendations have been re-stated and embodied in the various reports highlighted in Chapters 3 and 4, notably by Latham (1994), Egan (1998) and Comptroller and Auditor General (2001). However, unlike these latter reports which remain mainly a series of recommendations the British Property Federation system set out an approach to the delivery of its recommendations. One of the most radical proposals was that the 'Contractor's Schedule of Activities' should form the basis for tendering by the contractor and be used for the resource and financial management of the project. This approach was entirely sensible as it was using an accurate resource model (established by the contractor) not based on speculation or simulation as proposed in the Operational Bill described above. Like the Operational Bill it gained little support from the industry especially the design professionals and in particular, the quantity surveyors were hostile, as it recommended that the SMM based bill of quantities was not necessary within such a procurement strategy.

Reader reflection

It is recommended that if the readers seek further information on these proposals, and others that are less novel or radical, then they should study the various references highlighted in Chapters 2 and 5, as well as those mentioned above.

We have attempted to show, from first principles, why the need for resource-based models is necessary if we are to achieve effective financial management of the construction process from the earliest stages of design through the process of construction and the use of the building during its life. We have highlighted two earlier, radical approaches that were developed within the UK which showed that it was possible to produce such models. However, the uptake was extremely limited mainly due to conservatism and a culture of resistance within the construction industry at that time. Nonetheless, there is no doubt that these developments have influenced some of the improvements that have been incorporated in the current SMM, although such improvements have been minor and incremental in scope.

12.4   Information technology and classification

12.4.1   Classification

So what of the future for building design cost management? It would appear that all the levers for change are in place (as discussed in Chapters 3 and 4) and that lump sum, adversarially driven contracts based on financial competition are not recommended for current and future construction procurement. We have seen that design and build strategies incorporating partnership agreements, based on mutual trust and a sharing of ideas and information, can be a positive step forward to improve the poor performance of the industry.

So what is still required to allow this renaissance to take place and not lose momentum as has so often happened in the past? One solution is to improve the transparency of our information by the development of a comprehensive information system, underpinned by information technology that facilitates the accurate and rapid provision, manipulation and assembly of the specific information needs of all those concerned in the design, construction and operating of construction projects over their total life cycle as a basis for their effective and efficient management.

One can now analyse why even the provision of resource-based cost models, although steps in the right direction, are not on their own sufficient to provide effective cost management as clearly they do not satisfy the needs of our information system as defined above.

So what are the limitations of, for example, the Operational Bill or the British Property Federation system? It is simply that they are not part of a comprehensive, coordinated information system and thus are stand alone models reflecting a particular need. Hence they cannot, even with the aid of information technology, form part of a fully integrated system supplying relevant information to particular users at particular points in time in the design and construction process model.

Of course this argument does not just apply to our resource-based cost models but to all the various forms of project information used within the construction industry.

So how can such an integrated information system be achieved? Again, the answer is actually very simple but the implementation remains difficult. All that is required is that our information is organised such that all the data making up our information can be managed by the various providers and users of such information in terms of their provision, assembly and retrieval. This can be achieved by the classification or labelling of our data so that their whereabouts within the information system are always known, that their meaning is agreed and understood by all those involved in the management of design and construction, that their movement within the information system can be triggered and monitored and that their aggregation or disaggregation can be readily managed. Thus, if the data making up the database not only contain the resources that have or will create the construction project but also data describing the construction project in terms of its function or performance then we have a fully integrated information system.

Put very simply, the information system should fully describe the following at various stages during the maturation of the construction project:

- *What is it?*
- *How is it achieved?*
- *When is it achieved?*

Going back to our basic energy model developed at the beginning of this chapter, we can now see how it underpins the basis of our integrated information.

It all seems very simple and yet such an information system seems still a long way off. We have already made the point that the construction industry is full of inertia, protectionism and a reluctance to share information because of our historical evolution. We stated that classification or labelling is the essential requisite to allow us to organise our information system. What is a classification system? At its simplest level it is a mechanism to allow us to keep separate that which is different and keep together that which is the same. We use classification either formally or informally in our daily lives basically to allow us to decide where to store things, such as our clothes or house keeping information and of course to allow us to find what we are looking for later. We are sure that we have all experienced failures in our own personal classification systems especially first thing in the morning when we are short of time and cannot find what we are looking for. One of the biggest hurdles in designing an effective classification system is finding common agreement by all concerned as to what should be kept together and that which should be kept separate as different users have different priorities and interests. Considering the construction industry, one can quickly see that, for example, architects are much more interested in the performance of the construction project, such as the aesthetics and durability of the external wall, whereas to the manufacturer of the bricks in the external wall an entirely different set of criteria will be of interest, such as the clay, labour and plant associated with their manufacture. In other words, different users attach different attributes to the data making up their information needs.

Unfortunately information flow within the construction industry remains somewhat limited because of the general reluctance to develop and make use of an effective and embracing system of classification.

Not only is such a situation wasteful of time due to difficult search patterns but more importantly leads to duplication of effort and inevitable errors and mistakes occur as illustrated in Fig. 12.2. Figure 12.2 highlights a severe problem that the

Figure 12.2 Communication breakdowns in the construction industry (from an original cartoon by Christine Jaggar).

construction industry faces: poor and ineffective communications. We leave to the reader to work out which interests our characters represent.

The most successful and comprehensive classification system for the international building industry was the SfB faceted classification system which was initially developed by the pioneering Swedish architect Giertz in the 1940s. The system was felt to be so useful to the international construction community that it became a part of the work of the premier international research organisation the CIB. Details of the use and application of the SfB system with its three tables or facets can be found in the various reports produced by the CIB (CIB Working Commission W58–SfB Development Group 1973, CIB W74 Information Co-ordination for the Building Process 1986).

Arising out of the SfB system the CI/SfB system was developed primarily under the auspices of the RIBA (Ray-Jones and Clegg 1976). The approach, which was again a faceted approach, made use of the three SfB tables or facets but added two further facets; one reflecting the physical entity being produced (the building) and the other dealing with attributes relevant to but not directly relating to the project. (This particular facet included concepts such as standard forms, construction text books, research papers, etc.) The CI/SfB system became the UK standard for the classification of trade literature and was also adopted by the BCIS for the classification of the various building projects making up its database.

Dissatisfaction with the fragmented nature of the building industry came to a head in the late 1970s when, as mentioned in Chapter 3, coordinated project information was developed alongside the introduction of SMM 7. As explained in Chapter 3, this was the first time that the four major contributors to the design and construction process had actually come together to recognise the need to produce fully coordinated sets of project information within lump sum tendering strategies for the benefit of all concerned. As a result of their endeavours the Common Arrangement of Work Sections (CAWS), (CPI 1987b) was developed together with three codes of practice recommending how drawings, specifications and bills of quantities should be prepared to achieve fully coordinated and comprehensive project information. Unfortunately, as already highlighted in Chapter 3, despite the recommendations of the Latham report (1994), its use is still limited.

The latest approach to providing a classification system for the construction industry is Uniclass, which was referred to in Chapter 3. This approach is more comprehensive than all its predecessors in that it deals with two additional important areas of construction previously ignored. The first is that it includes the civil engineering industry and the committee responsible for Uniclass includes representatives from the Institution of Civil Engineers as well as the institutions representing the building industry as in the Co-ordinating Committee for Project Information (CPI 1987a). The second area which Uniclass addresses is that of the use of the construction project during its life as well as classifying the hardware making up construction projects, such as resources, elements and work sections, etc.

The aim of the system is to set out a definitive classification system for the construction industry by the use of its tables either separately or in combination for (Uniclass 1997):

■ arranging libraries
■ structuring product information

- coordinating project information
- structuring technical and cost information
- developing frameworks for databases.

In order that this can be achieved the Uniclass has developed the following tables:

A. Form of information

Used to organise reference materials in libraries, e.g. books, journals, CD-Roms, standards, legislation, dictionaries.

B. Subject disciplines

This table is used to organise information according to subject disciplines such as engineering, architecture, surveying, contracting, etc.

C. Management

This table comprises two sections (management and project management) and is designed to allow management and project management information to be classified. The project management section also allows information to be classified against the various stages in the project life cycle. This table covers areas such as management theory, corporate strategy, marketing, risk management, and project management.

D. Facilities

This table classifies construction work according to the activity or purpose for which it is intended. When used in combination with other tables it can be used to classify a construction complex, a construction entity or a space such as a hospital, prison, office, library or opera house.

E. Construction entities

A construction entity is an independent construction of a significant scale such as a building, a bridge or a dam. This table classifies construction entities according to physical form or basic function as opposed to user activity as in the previous table.

F. Spaces

This table allows for the classification of spaces according to a number of different characteristics including location, scale and degree of enclosure but not according to user activity such as rooms, court yards, public spaces.

G. Elements for buildings

Allows for the physical parts of buildings to be classified and is of course at the heart of design cost management as described in this book and demonstrated in the SFCA.

H. Elements for civil engineering works

This is intended to be used to aid the design cost management of civil engineering works and serves to complement Table G. Incidentally the BCIS have produced a Standard Form of Cost Analysis for Civil Engineering (CECA; Jaggar 1997), which is very similar to this particular table.

J. Work sections for buildings

This table is based on the CAWS as discussed above under the Coordinated Project Information initiative and is designed to be used to organise information to be contained in specifications and bills of quantities as discussed above.

K. Work sections for civil engineering works

This table is based on the CESMM3 (Institution of Civil Engineers 1985) and complements Table J. Typical examples are pipework, piling, and road surfacing.

L. Construction products

This is used for classifying trade literature and design/technical information relating to construction products. Typical examples are traffic signals, space frames lintels, bricks, office furniture and doors.

M. Construction aids

Used to classify trade literature and technical information according to plant and equipment used to aid construction operations, such as formwork, scaffolding, machines and tools.

N. Properties and characteristics

Used for classifying information on subjects relating to properties and characteristics and is intended to be used to arrange information in technical documents and for qualifying information contained in other tables, such as shape, size and appearance.

P. Materials

Used to classify different kinds of materials such as timber, cement, plastics and stone, and for further qualifying information from other tables especially Table L.

Q. Universal Decimal Classification (UDC)

This table indicates how UDC can be used to classify subjects not covered in the Uniclass system, such as philosophy, mathematics, sport, languages, geography, etc. This table provides the essential link between project specific information and general information: an essential requisite in information management.

As can be seen from this very brief description of Uniclass it attempts to provide a comprehensive classification system to allow the construction industry to organise its information as an essential prerequisite of any information management. The reader is advised to study the Uniclass document to gain a full understanding of faceted classification and the nature and purpose of the Uniclass system.

The development in the technologies to manage information has progressed quicker than the thinking in how to use the technology appropriately. The research community and practice have prioritised their efforts upon application of technology to manage information and the developments in coordinated project information using logical classification taxonomies have been largely abandoned. The information being managed by today's powerful technologies would be recognised as not having altered much by someone in practice 50 years ago and is still largely uncoordinated. However, technology will overcome a lot of the fragmentation and coordination issues once standards are developed. The authors believe that the technological advances should build upon the project information classification developments to ensure that the benefits of increased speed of information management are linked with the provision of more meaningful information that can be used in the cost modelling process. To gain a better idea of how technology is likely to push developments, an understanding of how information technology utilises object technology and how businesses are likely to communicate in the future is necessary.

12.5   Integration of information technology and design cost management processes

The foregoing chapters have illustrated how information is used by modelling techniques to provide advice to the design team at the various stages of design development. At the earliest stages the construction economist has very little specific design information to combine with the functional or per m^2 cost information. As the design develops the construction economist must use different models and information (see Fig. 12.3). The practices discussed do not allow for meaningful transfer of information from one model to the next. The group elemental cost model and the information supporting it becomes redundant as more detailed elemental design information becomes available that in turn becomes redundant as the bills of quantities are produced. This model redundancy recurs as the constructor generates other models that reflect the construction process, such as detailed programmes and resource allocation models, or method statements and unit rate allocations. These also become redundant once work commences upon site and accountancy-based models are used to monitor estimate accuracy against production costs.

The information supporting each one of these models could easily be stored in flat file databases; technology that was available 40 years ago. The models used in the practice of design cost management have not really progressed at all over the last 40 years. This could either be due to their being at a state of maturity that cannot be improved upon or more likely it is another instance of poor practice due to the fragmentation of the industry. The integrative technologies that can break through the translation barriers are now available and are discussed below.

The thinking underpinning a product model that supports differing ontological views of the product and process of construction was commenced over 60 years ago as outlined in Chapter 5. However, it is only now in the early twenty-first century that the construction industry has the technology to realise the vision of these early information management pioneers. Bindlsev encapsulated the philosophy of these models and the need for further research underpinning them in the following quotation:

> It is the task of administrative research to provide information systems that can help increase the quality of products within the building industry. It is

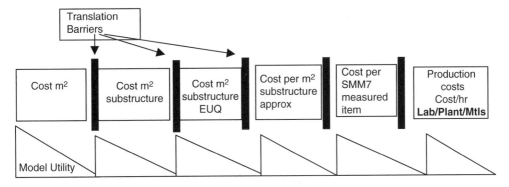

Figure 12.3   Model utility/translation barrier compared with design information availability.

a characteristic of building administration that it comprises the building project as a totality. ... It is important for rational achievement of the administrative processes that there is a logical connection between design information (drawings, specifications, bills of quantities, budgets, programmes) and recording information (measurements, descriptions, reports, accounts, etc.). ... It is impossible to control (a material, a construction, a finished product) unless the 'standards' by which control is facilitated have been recorded. These recorded 'standards' form the basis for comparison and subsequent action if the current observations or records are at variance.

(Bindslev 1995a, b)

Before looking in detail at the technology and how it can facilitate integration it is worth reflecting on the current capabilities of design and construction organisations to manage and exchange information. In a study by The Building Centre Trust (2001) of 400 individuals involved in 80 construction projects ranging from £10 million to £80 million the following was found:

■ Over 85% of the respondents to the survey had access to e-mail, internet and worked for an organisation that had its own network. However, only few projects of a significant value had a dedicated information technology network.

When the communicative use of the technology was investigated in further detail they found that at the design stage 91% of product specification information, 88% of cost information and 85% of subcontractor work package information was paper based. This dichotomy was also reflected at the construction stage to a similar level. For instance, 86% of the tender documents to main contractors, 85% tender documents to subcontractors and 88% of specifications as installed were in paper-based format.

Their survey concluded that the industry participants have the technology available to produce and use the documents. However, the barriers to moving away from the paper-based approaches are significant. These barriers are likely to be overcome in the near future as much of the development of the infrastructure for information system integration is already in place and is becoming more robust through significant developments in research and application in the manufacturing sector. We can consider infrastructure to include the information technology and communications hardware, software applications for design, communication and enterprise resource planning and significantly the development of agreed technical *interoperability* standards.

The benefits of information system integration are well known, the main ones are as follows:

■ Information sharing between the participants of processes.
■ Single point of data entry that increases speed and reduces error. This is particularly important when the design processes are considered and will be discussed later in this section.
■ Better quality control systems that can track changes to design or production information and will allow for post project audit and analysis.
■ Concurrent design processes whereby the parties can work on one design hub/file at the same time.

The first point of data entry into a project information system is generally at the design stage and a review of how the technologies have developed to support design processes is necessary in order to understand the potential role the construction economist may have in the future. Current computer-aided design (CAD) technologies are based upon database principles and allow for the storage of data in a range of formats; some visual such as lines, annotation text, rendered images and some calculated and a great deal of data that are not visual. The power of databases for the management of large bodies of data was realised in the early 1970s as following 'flat' file databases the relational format was developed. The relational format utilised a primary key relationship that allowed databases to be linked, which facilitated powerful data combinations and retrieval possibilities using the programmable structured query language (SQL). Database developers utilised entity relational modelling techniques to aid the design and structure of these systems that assisted with the normalisation and reduction of data redundancy, increased efficiency and improved the ease of maintenance.

The performance of relational databases in terms of information storage, retrievability and maintenance was poor compared to the newer object oriented databases developed in the late 1980s that utilised a fundamentally different means of organising data. These databases utilised object technology to organise data in terms of hierarchies based around classes and inheritance of attributes. This technology was utilised by CAD packages and is illustrated below.

If one considers the design of a wall in a CAD package, the physical representation of this element would simply be one or two lines that would have a thickness and two sets of coordinates, X and Y. If the design was in three dimensions a further coordinate, Z, may also be added. The opportunities available to the designer in using this technology to aid the design process are more than just drafting and visualising the result. The wall can be considered as an object in the design and as such can have a number of attributes associated with it. The *superclass* of walls might have attributes such as length, width, weight, acoustic performance, U value, etc. A *subclass* of walls might be internal walls that inherit all of the above attributes and also some attributes such as finish that are particular to the subclass. An instance of this class that would relate to the case study project would have a set of values that were project specific such as $21\,m$, $110\,mm$, $145\,kN/mm^2$, etc. The grouping of a set of *primitives* (lines) in a CAD can be used to develop objects that have three dimensions and can have a number of *attributes*. (BS1192, part 5, 1998 uses an example of a chair which has attributes of a seat, legs, back, etc.)

Attribute values are stored in object 'slots'. Slot values can be inputted by the designer as the design develops, can be automatically calculated from other design information or can be generated by other objects through the use of methods (a way of incorporating intelligence into objects via the use of *if…then* rules). This integration of intelligence in CAD is at a relatively low level of development at present. However, researchers with skills in artificial intelligence and construction are continually looking at how to capture, structure and incorporate object intelligence into designs. One can imagine that a set of rules could easily be stored for objects such as doors, for example, automatically check that the door met the relevant fire regulations, specified fixing into walls and even contained information regarding the resources required for fixing the object into position. This opportunity to attach attributes and their values to objects has been taken by some suppliers who provide designers with libraries of objects that can be incorporated into a design. These objects have all the relevant specification information structured and inputted by the component manufacturer. The need for robust standards is obvious as the designer may be faced with a set of incompatible file formats with information stored in a range of formats that may serve to confuse the design rather than augment it. This is being addressed by the International Agreement for Interoperability (IAI) standards which will be used to publish Industry Foundation Classes in the near future. These will provide an agreed protocol for describing building components.

Plainly if all the information was shown on a design drawing the resultant drawing would be unreadable or would need to be contained on a drawing larger that the maximum AO size currently used on sites in the UK. This difficulty is overcome by the CAD packages arranging the design information in a series of layers that can be considered as a series of transparent overlays. The drawings that are required for the steelwork frame fabricator would obviously be different than those for the floor finishes subcontractor. The benefit of CAD is that all this information can be contained on layered designs, the layers simply being turned on or off to match the supply chain parties' information requirements. Consequently, the designer's role becomes one of integrating different designs input via the use of a CAD package. The great benefits that this gives the designer are that all the design information is available on one file that can be integrated through the use of object oriented technology. The use of three-dimensional visualisation of the design can be used to detect design and detail clashes. The communication technology and interoperability standards development also makes concurrent design a possibility with a number of different parties working in *real time* on a design rather than sequentially as happens now. The reader is referred to the literature (Vincent 1995, ConstructIT 1999) for more information about the use of technology to integrate construction design processes and BS 1192 for details of the layering conventions.

The technological possibilities of incorporating cost information into designs has become more advanced with the introduction of CAD. To summarise, potential CAD systems can:

- Incorporate object cost information into a design that can be queried and used for cost modelling.
- Provide information that can be linked with other software applications such as enterprise resource planning systems, procurement systems and accounting.
- Will allow for concurrent design processes which includes dynamic updating of object details and resource programmes.

The processes of design cost management will be radically altered by these developments. Application of this technology will result in quicker design processes. The speedy exchange of more information will occur and less design (and on-site) reworking will be required. There is still a need for cost information systems that support this concurrent environment, dynamic results of *what if* scenarios, but building in intelligence of the cost implications of design alternatives should still be based on a quantitative analysis of cost drivers and their causal relationships. The practice of cost modelling can be more than speeded up and there now exists the possibility to radically change the methods and information that is used by the construction economist as the basis for judgement. The need for a logical classification system used to store relevant cost information, which may be attached to objects within the design is still a key challenge that has yet to be met.

Many barriers still exist to these developments and recent work by Fortune and Lees (1996) indicates that traditional practice still predominates within the industry. The reasons for designers having shown little interest in being responsible for the cost modelling of their designs were explored earlier in the text. However, the business environment in which design and construction is being undertaken is radically changing with the principles of lean construction, supply chain management and e-business approaches.

12.6   e-Business

There have been many advances made in the area of e-business and e-commerce in areas other than construction. Before we can consider the impact that these advances will have on the role and function of the construction economist it is worth defining the terms e-business and e-commerce in more detail.

e-Commerce tends to be the starting point for most organisations. It can be defined as the transaction of business using web technology between two parties without significant levels of integration occurring. Many of us have been involved at this level, whenever we buy a CD or airline ticket we simply buy a commodity or service only considering price, delivery date and reliability. e-Business however offers greater opportunities by integrating business processes and information.

e-Business aims to provide value to customers by achieving integration throughout the supply chain to allow for seamless flow of information. This information can be re-used in either the buyer's system; downstream or aggregated in the suppliers system, upstream. Plainly for e-business to work successfully an integrated information system, an agreed set of business processes and long term commitments are prerequisites. They exist in the car manufacturing sector in Merseyside and Jaguar cars recently expounded the benefits of e-business with their component suppliers. Another industry where it has been demonstrated to work is in the postal distribution service, whereby buyers of a service, that is delivery of a package, can view the current status of the service via the internet (whether it is in transit, in storage, delivered, etc.).

The software available to support e-business can be categorised into the following:

- Applications focused on the seller business processes that allow *shop-fronts* for products that can be customised depending upon the seller's relationship with the potential buyer of the product or service.
- Applications aimed at buyers to allow comparison and aggregation of products from approved suppliers, and the management of the procurement of the goods or services supplied, e.g. by automatic transmission of approved orders or comparison of subcontractor/supplier quotations.
- Applications that are focused on developing an open and accessible market. Buyers invite tenders on an open market or restricted market, the entry to which would be managed by meeting predefined criteria similar to selective tendering. The sellers of the service would be automatically informed of tendering

opportunities. This can also work in reverse whereby sellers with a product or service to sell can inform buyers of its availability, e.g. stockpiles of materials.

■ Applications that enable the integration of supply chains and assist the sellers in assessing the supply chains' capability.

■ Applications that integrate the above four applications and provide a mixture of legacy systems, enterprise resource planning (ERP) systems and customised application.

Most construction organisations have started on the road to e-business implementation with website development and emails being commonplace. The next exciting development will be the increasing use of some, or even all, of the applications outlined above in the construction industry. This will allow for more collaboration to exchange information and services in a controlled and managed manner. Tendering may become quicker and more open, market prices may become more competitive, the cost of governance of the project may also reduce significantly as communications are speeded up. The integration of the buy and sell sides of the supply chain may lead to a range of causal effects from one transaction downstream, invoicing, supply logistics, component assembly, stock transfer, labour and manufacturing plant scheduling could all flow on automatically. The reader is referred to ConstructIT's guide to e-business for further reading (ConstructIT for Business 2000).

Where does the construction economist fit into this integrated and automated world of common business processes, interoperability standards and intelligent CAD?

As long as construction projects require funding, clients and their design teams will always require an indication of cost implications of designs. The sources of cost information are likely to change as technology will provide supply chain participants the potential to provide, and communicate directly with the designer, budget estimates for their specialised elements of design at early stages. Whether they have the organisational capability to provide these estimates with integrity and reliability is an important question. One could argue that it is in their interest to develop this capability as it allows them to integrate at the earliest stages of the design process and consequently gain an advantage over their competitors. The concurrent design process could be extended to allow for automatic component prices to be available from builders merchant sources. This could be extended to more complex supply chain participants, such as steelwork subcontractors, who could provide an estimate for the supply and erection of a frame given certain conditions and this information may be directly available to the construction economist's computer as the design develops. One can imagine the CAD application steel frame design module having a set of criteria such as gross floor area, maximum span and building function that are inputted by the designer. The *buy side* application of an e-business system would search for a number of supply chain partners of steel frames (these would meet pre-specified criteria regarding their capability to carry out the work) who could provide a number of online budget estimates in real time. These could be compared with legacy records of similar projects' time, location, resources, etc. and provide a model for cost planning. The reliability of these estimates will still require assessment and monitoring against production standards. The construction economist's role will, in this scenario, become more sophisticated. With better quality and an increased quantity of information the construction economist will be able to request budget estimates. For example, for a portal steel frame for a two-storey production facility, work to be in a range of $\pm$ a given percentage at different stages of the design development. These standards of estimate would be assessed using a quantitative analysis of information from a range of projects that could incorporate more sophisticated information than the market price elemental unit rates which have been considered in the case study. The construction economist would be able to gather not only resource utilisation information from supply chain partners but would be able to gather information about the cost of components from builders merchants or raw materials directly from the steel rolling mills.

The supply chain participants' role in the design and construction will require management to ensure that the quality of the work packages are appropriate for the costs paid by the client. The construction economist will have to develop information systems that allow for the cost consequences of quality to be researched more objectively and to be able to then specify the quality of the product. Computer-based qualitative analysis tools are becoming available which have the ability to search and analyse large amounts of qualitative data and this may lead to a better understanding of the quality and semantics of the specification information used by design parties. Data analysis, data mining and knowledge discovery tools are becoming well established in other domains such as insurance, marketing and banking. These rely on large amounts of robust structured data to draw causal inferences, which may have the potential to be used in construction if the data and information are available. These may allow the construction economist the potential to provide the design team with advice on the cost consequences of using different alternatives, which would take account of the interdependencies of designs. Finer and more detailed information that is unencumbered with market prices would be essential to make such analysis meaningful.

The construction economist will still have to predict the costs of integration of the supply chain participants' contributions. This will, however, be a small proportion of the project's overall cost and at present this appears to be in the order of 10% of the costs of a project.

The above are merely conjectures about what may happen if the construction industry's supply chain management techniques, the integrative technologies and the motivation to integrate continue to develop. It is debatable whether the industry's clients are able to provide the stability of markets to allow for such developments to take place in anything less than 20 years. The UK government is, however, providing a strong lead with the encouragement of prime contracting and contractors, as discussed in Chapter 4, who work on a series of projects and will benefit from longer-term commitments.

The benefits of prime contracting and serial contractors:

■ greater investment in common systems for improving team working and information sharing
■ equitable and transparent sharing of the benefits of cost reduction
■ better sharing of information
■ getting suppliers' input into design
■ getting contractors' and suppliers' input into design for improved constructability.

The key issue is to transform business processes to exploit the new tools available. This involves firms adapting their in-house information technology systems and moving towards mutually compatible business processes. The only way this can happen is through the development and adoption of industry standards. These may allow for greater levels of integration. The construction industry is addressing the issue of organisation's capability by considering processes and have started research into the development of a set of protocols for the development of mature processes (Aouad *et al.* 1999, Sarshar *et al.* 1999).

The barriers to the integration of construction industry information systems are strong. The Building Centre report on the effective integration of information technology in construction (Goodwin 2000) identifies the following barriers:

- uncertainty at project inception
- shared power structure on majority of projects
- low levels of support for integration within the project team
- numerous sources of inertia from participating organisations.

However, there are equally strong forces at work to break down these barriers:

- better briefing processes
- integrated methods of procurement
- better skills in the use of information technology and better understanding of its benefits
- supply chain management principles encouraging integration between organisations if they are to survive.

Information standards are at the heart of any project information strategy. This is not just capturing in electronic form what is currently done on paper, but should be developed to consider the benefits that may accrue in innovative forms of data and information combinations. There are many barriers to integration (as have been discussed) but the move towards a partnering environment with long term non-adversarial relationships between the parties is a good step along the way to reducing the fragmentation barriers. The use of framework contracts that ensure continuity of work from major clients puts in place the conditions supporting the investment in information technology systems by parties who have a vested interest in effective communication and information exchange. This overcomes the current arrangements whereby lots of organisations have their own methods for information format and exchange. The technology that is available to the firms is becoming more 'open'; e-mail standards, word processing and html files are all used extensively. However, to yield the greater benefits of integration it must be deeper than merely using the technology to pass messages that previously existed in paper form.

12.7   Conclusion

The technological enablers to radically change the role of the construction economist are already available. The change in construction procurement provides a real opportunity to make supply chain knowledge more available to designers and constructors. There is a need to break down the economic barriers between the participants, disentangle production cost information from market price information and provide an information system that can pull the various sources of knowledge together. What needs to be avoided is the creation of even more black boxes of cost information that have electronic as well as organisational barriers around them. The authors suggest that at the heart of such an information system should be the construction programme for the particular construction project as it reflects the resources required, when they are required, what is to be achieved: the completed construction project. It goes without saying that such a programme should be priced by the constructors in order to form the basis of financial management for the project and the cost information.

To ensure the success of such an approach it is necessary for the construction contracts to be used as the basis for setting the contractual arrangements for the particular procurement strategy to be deployed to set out clearly the requirements of the information system to be used for the construction project.

Typical clauses might be:

■ The basis of resource and financial management for the project shall be the constructor's priced programme which shall indicate the activities to be carried out, their sequence, the resources needed for their completion and their costs. This programme is to be a contract document and is to be agreed to by all parties involved before commencement of any construction work. The resources required shall not be subject to change in quantity, utilisation or price unless design changes are introduced.
■ The information contained in the various activities, together with the resources required, shall be classified in accordance with the tables of *Uniclass* in terms of:

(1) G. Elements for buildings and/or H. Elements for civil engineering works
(2) J. Work sections for buildings and/or K. Work sections for civil engineering works
(3) L. Construction products
(4) M. Construction aids
(5) P. Materials.

The future for project information with the development of e-commerce and the internet as demonstrated by the website accompanying this book, is clear, in that printed documentation will be required less and less and users will be able to obtain and expect to be able to carry out specific information searches and retrievals whenever it is required as shown in Fig. 12.4.

We have shown that for design cost management and the construction industry at large to improve its performance in terms of quality, cost and time as required by our clients and highlighted in the book, there are two essential prerequisites:

(1) The effective identification and management of construction resources
(2) The development of an effective information management system based on classification or labelling of our construction information.

We have explained and demonstrated how this might be done and highlighted the mechanisms for its successful implementation. As discussed in Chapter 4,

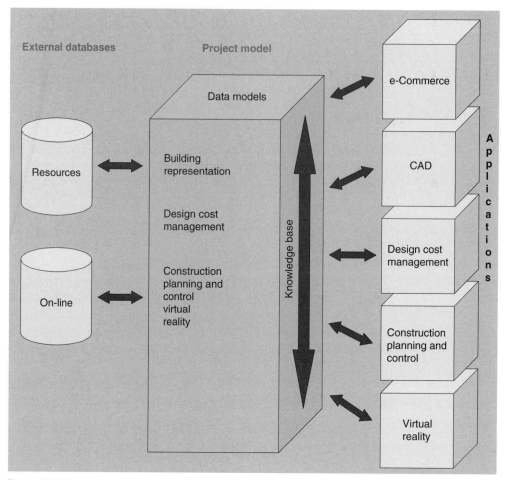

Figure 12.4   Overview of project information system. Adapted from Space project, Salford University.

the industry at large led by the leading client bodies has lost patience with the construction industry and its inefficiencies, and are encouraging and recommending the use of more open and harmonious procurement strategies based on the use of design and build incorporating partnership arrangements. The time has never been better for the industry to rethink what it does which was aptly summed up by Egan when he stated:

> The task force wishes to emphasise that they are not inviting the UK construction industry to look at what it already does and *do it better*. They are asking the industry and Government to join with major clients to *do it entirely differently*.
>
> (Egan 1998)

12.8   Reader reflections

■ How will object technologies influence the design processes and consequently the design cost management processes?

■ How will information technology and web-based communication technologies influence procurement systems in the future?

■ Information without standardised classification makes storage, searching, manipulation and communication very difficult. Imagine how difficult it would be to find this book in the library or the webpage on the internet without a commonly recognised system. What are the reasons for the stalling of the developments in coordinated project information?

■ The authors advocate a move to process-based rather than product-based procurement documentation. What barriers must be overcome if these recommendations are to be adopted?

■ What do you consider to be a realistic assessment of the time required for fundamental changes, as suggested in this chapter, to take place in the UK construction industry?

A1 Web information

The website address is www.bdcm.co.uk

The material on the website is constantly updated however at the date of publication the following information is available.

Structure of textbook

- Chapter headings
- Fact files
- Reader reflections

Case study information

- client's brief
- site layout   drawing no. 3250/04
- south elevation
- east elevation
- north elevation
- west elevation
- floor plans   drwg no. 3250/11
- foundation proposal   drwg no. 23648/p100
- floor layout   drwg no. 23648/p101
- roof layout   drwg no. 23648/p102
- site investigation report
  and borehole information
- photographs of the
  completed building

BCIS general information

BCIS Standard form of cost analysis
BCIS Standard elements
BCIS Detailed form of cost analysis

BCIS cost information

Average building prices for factories
Group elemental cost analysis for factory A
Detailed elemental cost analyses for factories B, C, D and E
Relevant tender price index and regional index information for the above analyses and the case study in Liverpool

Cost models

Outline cost plan modelling tool
Detailed elemental cost plan tool

Outputs

Cost bracket
Outline cost plan
Detailed elemental cost plan
Elemental cost checks for all elements

Other information

Useful links
Information to assist the teaching and learning of Design Cost Management
About the authors
Practice profiles

A2 Cost information

BCIS Standard form of cost analysis
BCIS Standard elements
BCIS Detailed form of cost analysis

Standard Form
of
Cost Analysis

PRINCIPLES, INSTRUCTIONS AND DEFINITIONS

The Building Cost Information Service
of
The Royal Institution of Chartered Surveyors,
12 Great George Street,
Parliament Square,
London SW1P 3AD.

December 1969
(Reprinted August 2001)

© 1969 The Royal Institution of Chartered Surveyors
All rights reserved
85406

CONTENTS

INTRODUCTION

The purpose of cost analysis is to provide data which allows comparisons to be made between the cost of achieving various building functions in one project with that of achieving equivalent functions in other projects. It is the analysis of the cost of a building in terms of its elements. An element for cost analysis purposes is defined as a component that fulfils a specific function or functions irrespective of its design, specification or construction. The list of elements, however, is a compromise between this definition and what is considered practical.

The cost analysis allows for varying degrees of detail related to the design process; broad costs are needed during the initial period and progressively more detail is required as the design is developed. The elemental costs are related to square metre of gross internal floor area and also to a parameter more closely identifiable with the elements function, i.e. the element's unit quantity. More detailed analysis relates costs to form of construction within the element shown by 'All-in' unit rates.

Supporting information on contract, design/shape and market factors are defined so that the costs analysed can be fully understood.

The aim has been to produce standardisation of cost analyses and a single format for presentation.

This document has been prepared jointly by J.D.M.Robertson, F.R.I.C.S., A.M.B.I.M., on behalf of the R.I.C.S. Building Cost Information Service and by R.S.Mitchell, A.R.I.C.S., on behalf of the Ministry of Public Building and Works. The principles and definitions are based upon the report of a working party under the chairmanship of E.H.Wilson, F.R.I.C.S., and the analysis of services elements has had the assistance of a report by a working party under the chairmanship of A.W.Ovenden, F.R.I.C.S.

The principles and definitions of cost analysis and this format are supported by:-

> The Quantity Surveyor's Committee of The Royal Institution of Chartered Surveyors
>
> The R.I.C.S. Building Cost Information Service, and
>
> The Chief Quantity Surveyors of:-
>> The Ministry of Public Building and Works.
>>
>> The Department of Health and Social Security.
>>
>> The Ministry of Housing and Local Government.
>>
>> The Department of Education and Science.
>>
>> The Home Office.
>>
>> The Scottish Development Department.

THE STANDARD FORM OF COST ANALYSIS

1: PRINCIPLES OF ANALYSIS

The basic principles for the analysis of the cost of building work are as follows:-

1.1 A building within a project shall be analysed separately.

1.2 Information shall be provided to facilitate the preparation of estimates based on abbreviated measurements.

1.3 Analysis shall be in stages with each stage giving progressively more detail; the detailed costs at each stage should equal the costs of the relevant group in the preceding stage. At any stage of analysis any significant cost items that are important to a proper and more useful understanding of the analysis shall be identified.

1.4 Preliminaries shall be dealt with as prescribed for the appropriate analyses.

1.5 Lump sum adjustments shall be spread pro-rata amongst all elements of the building(s) and external works based on all work excluding Prime Cost and Provisional Sums contained within the elements.

1.6 Professional fees shall not form part of the cost analysis.

1.7 Contingency sums to cover unforeseen expenditure shall not be included in the analysis of prices, but shown separately.

1.8 The principal cost unit for all elements of the building(s) shall be expressed in £ to two decimal places per square metre of gross internal floor area.

1.9 A functional unit cost shall be given.

1.10 In Amplified Analyses, design criteria shall be given against each element. Special design and performance problems shall be identified.

1.11 The definitions of terms for cost analysis shall be those given hereafter.

1.12 The elements for cost analysis shall be those given hereafter.

1.13 The contents of each element shall be as given hereafter.

1.14 The principles of further detailed analysis shall be as given hereafter.

2: INSTRUCTIONS

2.1 GENERALLY

2.1.1 Definition of terms
Definitions of terms used throughout the analysis follow these instructions.

2.1.2 Complex contracts
A cost analysis must apply to a single building. In a complex contract (i.e. a contract which contains a requirement for the erection of more than one building) the size of the contract may have an important bearing on price levels obtained. If this situation occurs, it should be identified in the box "Project details and site conditions" on the first sheet of the analysis.

2.1.3 Omissions or exclusions
Where items of work which are normally provided under the building contract have been excluded or supplied separately, this should be stated where appropriate.

2.2 PROJECT INFORMATION

2.2.1 Building type
CI/SfB classification will be given and restricted to the "Built environment" classification taken from Table 0.

A "College of further education" will therefore be classified and shown as:-

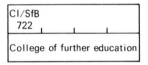

```
CI/SfB
 722
College of further education
```

2.2.2 BCIS code
The BCIS reference code classifies buildings by the form of construction, number of storeys and gross internal floor area in square metres.

The different construction classes are:-

A   Steel framed construction

B   Reinforced concrete framed construction

C   Brick construction

D   Light framed steel or reinforced concrete construction.

A single-storey building of 766 square metres of gross internal floor area and built in traditional construction would have the following BCIS code:-   C - 1 - 766.

2.2.3 Client
Indication should be given of the type of client, e.g. borough council; church authority; owner-occupier; government department; property company; etc.

2.2.4 Location
The location of the project should be given, noting the city or the county borough or alternatively the borough and the county, e.g. Bristol, or Richmond, Surrey. The location may be reported less precisely if the client so desires.

2.2.5 Tender date
(1) Date fixed for receipt of tenders.
(2) "The date of tender" i.e. 10 days before date of receipt.

2.2.6 Brief description of total project

(a) Brief description of the building being analysed and of the total project of which it forms part.

(b) Any special or unusual features affecting the overall cost not otherwise shown or detailed in the analysis.

2.2.7 Site conditions

(a) Site conditions with regard to access, proximity of other buildings and construction difficulties related to topographical, geological or climatic conditions.

(b) Site conditions prior to building, e.g. woodland, existing building, etc.

2.2.8 Market conditions

Short report on tenders indicating the level of tendering, local conditions with regard to availability of labour and materials, keenness and competition.

2.2.9 Contract particulars of total project

(a) Type of contract, e.g. J.C.T. (with or without Quantities), GC/Wks/1, etc.

(b) Bills of quantities, bills of approximate quantities, schedule of rates.

(c) Open or selected competition, negotiated, serial or continuation contract.

(d) Firm price or, if fluctuating, whether for labour or materials or both.

(e) Number of contractors to which tender documents sent.

(f) Number of tenders received.

(g) Contract periods: (i) stipulated by client;
(ii) quoted by builder.

2.2.10 Tender list

(a) List of tenders received in descending value order.

(b) Indicate whether tenders were from local builders (L) or builders acting on a national scale (N).

2.3 DESIGN/SHAPE INFORMATION OF SINGLE BUILDINGS

2.3.1 Accommodation, design features

(a) General description of accommodation.

(b) Where a building incorporates more than one function (e.g. a block of offices with shops or car park deck) the gross floor areas of each should be shown separately.

(c) Where drawings are not provided, a thumbnail sketch shall be given of the building showing overall dimensions and number of storeys in height for each part related to ground floor datum (i.e. $\oplus$ for ground floor and upper storeys, $\ominus$ for basement storeys).

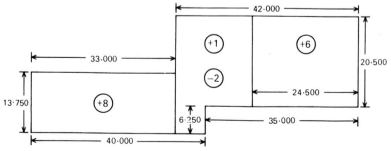

(d) Any particular factors affecting design/cost relationship resulting from user requirements or dictates of the site (user requirement is defined in R.I.B.A. Handbook as the area of accommodation, the activities for which a building is required, and the quality and standards it should achieve as stated by the client).

2.3.2 Floor areas
The measurement of floor areas and the information required on the form is that detailed hereafter under "Definitions".

2.3.3 Number of storeys
(a) Approximate percentage of building (based on gross floor area) having different number of storeys, i.e. 20% single storey, 30% two storey, 50% three storey.

(b) Excludes structures such as lift, plant or tank rooms and the like above main roof slab.

2.3.4 Storey height
Storey heights shall be given and differing heights stated separately.   (See Definitions.)

2.4 BRIEF COST INFORMATION

2.4.1 Preliminaries and contingencies
The totals for preliminaries and contingencies for the building being analysed should be stated and also each should be expressed as a percentage of the remainder of the contract sum. The analysis of prices does not include Contingencies.

2.4.2 Functional unit cost
The functional unit cost should be calculated by dividing the total of the group elements (including preliminaries but excluding external works) by the total number of functional units. (See Definitions.)

2.5 SUMMARY OF ELEMENT COSTS

2.5.1 Elements
The building prices are analysed by the elements.

Preliminaries are shown separately and also apportioned amongst the elements.

Where Preliminaries are shown separately each element is analysed under the following headings:-

(a) Total cost of element.

(b) Cost per square metre of gross floor area £ and decimal parts of a £ (to two decimal places).

(c) Element unit quantity (as later described).

(d) Element unit rate £ and decimal parts of a £ (to two decimal places).

Where Preliminaries have been apportioned each element is analysed under the following headings:-

(a) Total cost of element.

(b) Cost per square metre of gross floor area £ and decimal parts of a £ (to two decimal places).

2.5.2 Group elements
Sub-totals are shown for the group elements:- substructure, superstructure, internal finishes, fittings and furnishings, services and external works.

Costs per square metre of gross floor area expressed in £ and decimal parts of a £ are calculated for the group elements.

Costs per square metre are also shown adjusted to a base date and in the case of analyses submitted to BCIS this will be made by the Service using the BCIS cost indices.

2.6 AMPLIFIED COST ANALYSIS

2.6.1 Elements
The standard list of elements to be used for amplified cost analyses is that described in the following pages. The cost of each element must conform with the appropriate list of items shown in the specification notes for each element and with the principles of analysis.

2.6.2 Design criteria and specification
Design criteria relate to requirements, purpose and function of the element, and an outline of the design criteria is noted under each element.

The specification notes are considered to reflect architects' solution to the conditions expressed by the design criteria and should indicate the quality of building achieved.

Specification notes provide a check list of the items which should be included with each elements. Notes should adequately describe the form of construction and quality of material sufficiently to explain the costs in the analysis.

The instructions of the specification notes which follow are definitions of principle and where any departure from them seems necessary, a note should be made explaining how these cases have been dealt with.

2.6.3 Preliminaries
In the Amplified Analysis the element costs do not include Preliminaries which are to be analysed separately. However, under each element is to be included a figure which represents Preliminaries expressed as a percentage of the remainder of the contract sum.

2.6.4 External works
The expression of cost of external works related to gross floor area of the building(s) is not particularly meaningful. It is used in the "Summary of element costs" so that the totals agree arithmetically, but in the Amplified Analysis there are no detailed costs required by this method.

2.6.5 Total cost of element
This is the cost of each element and the items comprising it should correspond with the notes in the right-hand column headed "Specification".

If no cost is attributed to an element, a dash should be inserted in the cost column and a note made to the effect that this element is not applicable.

Where the costs of more than one element are grouped together, a note should be inserted against each of the affected elements, explaining where the costs have been included. For example, if windows in curtain walling are included in "External walls" it should be so stated in the element "Windows" and details of the cost included with the element "External walls".

2.6.6 Cost of element per square metre of gross floor area
This is the "Total cost of element" divided by the gross floor area of the building.

2.6.7 Element unit quantity
In an amplified analysis, the cost of the element is expressed in suitable units which relate solely to the quantity of the element itself.

Instructions are given in the appropriate column which show what element unit quantity is to be used for each element, e.g. in the case of "Floor finishes" the element unit quantity is the "Total area of the floor finishes in square metres" and in the case of "Heat source" the element unit quantity is "kilowatts".

(i) Area for element unit quantities

All areas must be the net area of the element, e.g. external walls should exclude window and door openings, etc.

(ii) Cubes for element unit quantities

Cubes for air conditioning, etc., shall be measured as the net floor area of that part treated, multiplied by the height from the floor finish to the underside of the ceiling finish (abbreviated to Tm^3).

2.6.8   Element unit rate

This is the total cost of the element divided by the element unit quantity. In effect it includes the main items and the labour items of the element expressed in terms of that element's own parameter. For example, in the case of "Floor finishes" it is the total cost of the floor finishes divided by their net areas in square metres; in the case of "Heat source" the elemental unit rate is the total cost of the heat source divided by its own parameter, the number of kilowatts. Elemental unit rates are shown in £ and decimal parts of a £ (to two decimal places).

2.6.9   Further quality breakdown

Where various forms of construction or finish exist within one element the net areas and costs of the various types of construction should be included separately in the specification notes and provision has been made where appropriate.

The area of each form of construction is the net area involved and excludes all openings, etc.

The cost of each form of construction is the total cost of all items pertaining to that construction.

2.6.10 The following is an example of how the amplified analysis form should be completed.

Element of design criteria	Total cost of element £	Cost of element per m^2 of gross floor area £	Element unit quantity	Element unit rate £	Specification
3.B Floor finishes	1,664	2.17	694 m2	2.40	19 mm granolithic laid monolithic, no skirting. 3 mm thermoplastic tiles Series 2 on 48 mm cement and sand screed, softwood skirting. 3 mm vinylised tiles on 48 mm cement and sand screed, softwood skirting. 25 mm (1") "West African" sapele wood block floor on 37 mm cement and sand screed, softwood skirting. 16 mm (⅝") red quarries on 34 mm screed, quarry skirting.

Floor finishes	£	Area m2	All-in unit rate £
19 mm granolithic	30	30	1.00
3 mm thermoplastic Series 2	18	13	1.39
3 mm vinylised tiles	616	395	1.56
25 mm (1") sapele blocks	570	161	3.54
16 mm (⅝") quarries	430	95	4.52

Preliminaries 9.73% of remainder of contract sum.

3: DEFINITIONS

3.1 Enclosed spaces

1. All spaces which have a floor and ceiling and enclosing walls on all sides for the full or partial height.

2. Open balustrades, louvres, screens and the like shall be deemed to be enclosing walls.

3.2 Basement floors

All floors below the ground floor.

3.3 Ground floor

The floor which is nearest the level of the outside ground.

3.4 Upper floors

All floors which do not fall into any of the previously defined categories.

3.5 Gross floor area

1. Total of all enclosed spaces fulfilling the functional requirements of the building measured to the internal structural face of the enclosing walls.

2. Includes area occupied by partitions, columns, chinmey breasts, internal structural or party walls, stairwells, lift wells, and the like.

3. Includes lift, plant, tank rooms and the like above main roof slab.

4. Sloping surfaces such as staircases, galleries, tiered terraces and the like should be measured flat on plan.

    Note: (i) Excludes any spaces fulfilling the functional requirements of the building which are not enclosed spaces (e.g. open ground floors, open covered ways and the like). These should each be shown separately.

    (ii) Excludes private balconies and private verandahs which should be shown separately.

3.6 Net floor area

Net floor area shall be measured within the structural face of the enclosing walls as "Usable", "Circulation" and "Ancillary" as defined below. Areas occupied by partitions, columns, chimney breast, internal structural or party walls are excluded from these groups, and are shown separately under "Internal divisions".

1. **Usable**

    Total area of all enclosed spaces fulfilling the main functional requirements of the building (e.g. office space, shop space, public house drinking area, etc.).

2. **Circulation**

    Total area of all enclosed spaces forming entrance halls, corridors, staircases, lift wells, connecting links and the like.

3. **Ancillary**

    Total area of all enclosed spaces for lavatories, cloakrooms, kitchens, cleaners' rooms, lift, plant and tank rooms and the like, supplementary to the main function of the building.

4. **Internal divisions**

    The area occupied by partitions, columns, chimney breasts, internal structural or party walls.

    Note: The sum of the areas falling in the categories defined above will equal the gross floor area.

3.7 Net habitable floor area (residential buildings only)

1. Total area of all enclosed spaces forming the dwelling measured within the structural internal face of the enclosing walls.

2. Includes areas occupied by partitions, columns, chimney breasts and the like.

3. Excludes all balconies, public access spaces, communal laundries, drying rooms, lift, plant and tank rooms and the like.

3.8 Roof area

1. Plan area measured across the eaves overhang or to the inner face of parapet walls.

2. Includes area covered by rooflights.

3. Sloping and pitched roofs should be measured on plan area.

3.9 External wall area

The wall area of all the enclosed spaces fulfilling the functional requirements of the building measured on the outer face of external walls and overall windows and doors, etc.

3.10 Wall to floor ratio

Calculated by dividing the external wall area by the gross floor area to three decimal places.

3.11 Element ratios

Calculated by dividing the net area of the element by the gross floor area to three decimal places.

Note: In the case of buildings, where only a part is treated or served by mechanical or electrical installations, indication of this is given by a ratio as follows:-

$$\frac{t \; m^2}{\text{gross floor area}}$$

where $t \; m^2$ is the total net area in square metres of the various compartments treated or served.

3.12 Storey height

1. Height measured from floor finish to floor finish.

2. For single-storey buildings and top floor of multi-storied buildings, the height shall be measured from floor finish to underside of ceiling finish.

3.13 Internal cube

1. To include all enclosed spaces fulfilling the requirements of the building.

2. The cube should be measured as the gross internal floor area of each floor multiplied by its storey height.

3. Any spaces fulfilling a requirement of the building, which are not enclosed spaces, such as open ground floors, open covered ways and the like, should be shown separately giving the notional cubic content of each, ascertained by notionally enclosing the open top or sides.

3.14 Functional unit

The functional unit shall be expressed as net usable floor area (offices, factories, public houses, etc.) or as a number of units of accommodation (seats in churches, school places, persons per dwelling, etc.).

4: FORMS OF ANALYSIS AND GUIDANCE NOTES

4.1 Concise, detailed and amplified forms of cost analysis

The standard method of cost analysis described here is in stages with each stage giving progressively more detail; the detailed costs at each stage should equal the costs of the relevant group in the preceding stage. At any stage of analysis any significant cost items that are important to a proper and more useful understanding of the analysis should be identified.

Forms of cost analysis have been prepared in three degrees of detail, Concise, Detailed and Amplified. The Detailed and Amplified forms are laid out as follows:-

> General and Background information and Summary of Elemental Costs for use with Detailed and Amplified Analyses.

> Check list of Specification and Design Notes which should accompany the Detailed Analysis.

> Breakdown of the information required in an Amplified Analysis.

For details of the Concise Cost Analysis see The Building Cost Information Service Section H.

4.2 Tank rooms

Where tank rooms, housing and the like are included in the gross floor area, their component parts shall be analysed in detail under the appropriate elements. Where this is not the case, their costs should be included as "Builder's work in connection" (5.14).

4.3 Glazing and ironmongery

Glazing and ironmongery should be included in the elements containing the items to which they are fixed.

4.4 Decoration

Decoration, except to fair-faced work, should be included with the surface to which it is applied, allocated to the appropriate element, and the costs shown separately. Painting and decorating to fair-faced work is to be treated as a "Finishing".

4.5 Chimneys

Chimneys and flues which are an integral part of the structure shall be included with the appropriate structural elements.

4.6 Drawings

Drawings, A4 size negatives if these are available, should preferably accompany Detailed Cost Analyses.

BCIS Standard Elements

1 SUBSTRUCTURE
All work below underside of screed or where no screed exists to underside of lowest floor finish including damp-proof membrane, together with relevant excavations and foundations.

2 SUPERSTRUCTURE
2A Frame - Loadbearing framework of concrete, steel or timber. Main floor and roof beams, ties and roof trusses of framed buildings. Casing to stanchions and beams for structural or protective purposes.
2B Upper Floors - Upper floors, continuous access floors, balconies and structural screeds suspended floors over or in basements.
2C Roof - Roof Structure - Construction, including eaves and verges, plates and ceiling joists, gable ends, internal walls and chimneys above plate level, parapet walls and balustrades. Roof Coverings - Roof screeds and finishings. Battening, felt, slating, tiling and the like. Flashings and trims. Insulation. Eaves and verge treatment. Roof drainage - Gutters where not integral with roof structure, rainwater heads and roof outlets. (Rainwater downpipes to be included in 'Internal Drainage' (5C). Roof Lights - Roof lights, opening gear, frame, kerb and glazing. Pavement lights.
2D Stairs - Stair Structure - Construction of ramps, stairs and landings other than at floor levels. Ladders. Escape staircases. Stair Finishes - Finishes to treads, risers, landings (other than at floor levels), ramp surfaces, strings and soffits. Stair Balustrades and Handrails - Balustrades and handrails to stairs, landings and stairwells.
2E External Walls - External enclosing walls including that to basements but excluding items included with 'Roof Structure' (2C). Chimneys forming part of external walls up to plate level. Curtain walling, sheeting rails and cladding. Vertical tanking. Insulation. Applied external finishes.
2F Windows and External Doors - Windows - Sashes, frames, linings and trims. Ironmongery and glazing. Shop fronts. Lintels, sills, cavity damp-proof courses and work to reveals of openings. External Doors - Doors, fanlights and sidelights. Frames, linings and trims. Ironmongery and glazing. Lintels, thresholds, cavity damp-proof courses and work to reveals of openings.
2G Internal Walls and Partitions - Internal walls, partitions and insulation. Chimneys forming part of internal walls up to plate level. Screens, borrowed lights and glazing. Moveable space-dividing partitions. Internal balustrades excluding items included with 'Stair balustrades and handrails' (2D).
2H Internal Doors - Doors, fanlights and sidelights. Sliding and folding doors. Hatches. Frames, linings and trims. Ironmongery and glazing. Lintels, thresholds and work to reveals of openings.

3 FINISHES
3A Wall Finishes - Preparatory work and finishes to surfaces of walls internally. Picture, dado and similar rails.
3B Floor Finishes - Preparatory work, screeds, skirtings and finishes to floor surfaces excluding items included with 'Stair Finishes' (2D) and structural screeds included with 'Upper Floors' (2B). Access Floors.
3C Ceiling Finishes - Finishes to Ceilings - Preparatory work and finishes to surfaces of soffits excluding items included with 'Stair Finishes' (2D) but including sides and soffits of beams not forming part of a wall suface. Cornices, coves. Suspended Ceilings - Construction and finishes of suspended ceilings.

4 FITTINGS AND FURNISHINGS
4A Fittings and Furnishings - Fittings, Fixtures and Furniture - Fixed and loose fittings and furniture including shelving, cupboards, wardrobes, benches, seating, counters and the like. Blinds, blind boxes, curtain tracks and pelmets. Blackboards, pin-up boards, notice boards, signs, lettering, mirrors and the like. Ironmongery. Soft Furnishings - Curtains, loose carpets or similar soft furnishing materials. Works of Art - Works of art if not included in a finishes element or elsewhere. Equipment - Non-mechanical and non-electrical equipment related to the function or need of the building (eg gymnasia equipment).

5 SERVICES
5A Sanitary Appliances - Baths, basins, sinks etc. WC's, slop sinks, urinals and the like. Toilet-roll holders, towel rails etc. Traps, waste fittings, overflows and taps as appropriate.
5B Services Equipment - Kitchen, laundry, hospital and dental equipment and other specialist mechanical and electrical equipment related to the function of the building.
5C Disposal Installations - Internal Drainage - Waste pipes to 'Sanitary appliances' (5A) and 'Services Equipment' (5B). Soil, anti-syphonage and ventilation pipes. Rainwater downpipes. Floor channels and gratings and drains in ground within buildings up to external face of external walls. Refuse Disposal - Refuse ducts, waste disposal (grinding) units, chutes and bins. Local incinerators and flues thereto. Paper shredders and incinerators.
5D Water Installations - Mains supply - Incoming water main from external face of external wall at point of entry into building including valves, water meters, rising main to (but excluding) storage tanks and main taps. Insulation. Cold Water Services - Storage tanks, pumps, pressure boosters, distribution pipework to sanitary appliances and to services equipment. Valves and taps not included with 'Sanitary appliances' (5A) and/or 'Services equipment'(5B). Insulation. Hot Water Services - Hot water and/or mixed water services. Storage cylinders, pumps, calorifiers, instantaneous water heaters, distribution pipework to sanitary appliances and services equipment. Valves and taps not included with 'Sanitary appliances' (5A) and/or' Services equipment' (5B). Insulation. Steam and Condensate - Steam distribution and condensate return pipework to and from services equipment within the building including all valves, fittings etc. Insulation.
5E Heat Source - Boilers, mounting, firing equipment, pressurising equipment, instrumentation and control, ID and FD fans, gantries, flue and chimneys, fuel conveyors and calorifiers. Cold and treated water supplies and tanks, fuel oil and/or gas supplies, storage tanks etc, pipework, (water or steam mains) pumps, valves and other equipment. Insulation.

5F Space Heating and Air Treatment - Water and/or Steam (Heating only) - Heat emission units (radiators, pipe coils etc) valves and fittings, instrumentation and control and distribution pipework from 'Heat source' (5E). Ducted Warm Air (Heating only) - Ductwork, grilles, fans, filters etc, instrumentation and control. Electricity (Heating only) - Cable heating systems, off-peak heating system, including storage radiators. Local Heating (Heating only) - Fireplaces (except flues), radiant heaters, small electrical or gas appliances etc. Other Heating Systems (Heating only). Heating with ventilation (Air treated locally) - Distribution pipework ducting, grilles, heat emission units including heating calorifiers, except those which are part of 'Heat source' (5E) instrumentation and control. Heating with Ventilation (Air treated centrally) - All work as detailed above for system where air treated centrally. Heating with Cooling (Air treated locally) - All work as detailed above including chilled water systems and/or cold or treated water feeds. The whole of the costs of the cooling plant and distribution pipework to local cooling units shall be shown separately. Heating with Cooling (Air treated centrally) - All work detailed above for system where air treated centrally.
5G Ventilating System - Mechanical ventilating system not incorporating heating or cooling installations including dust and fume extraction and fresh air injection, unit extract fans, rotating ventilators and instrumentation and controls.
5H Electrical Installations - Electric Source and Mains - All work from external face of building up to and including local distribution boards including main switchgear, main and sub-main cables, control gear, power factor correction equipment, stand-by equipment, earthing etc. Electric Power Supplies - All wiring, cables, conduits, switches from local distribution boards etc, to and including outlet points for individual installations. Electric Lighting - All wiring, cables, conduits, switches etc from local distribution boards and fittings to and including outlet points. Electric Light Fittings - Light fittings including fixing. 5I Gas Installations - Town and natural gas services from meter or from point of entry where there is no individual meter: distribution pipework to appliances and equipment.
5J Lift and Conveyor Installations - Lifts and Hoists - The complete installation including gantries, trolleys, blocks, hooks and ropes, downshop leads, pendant controls and electrical work from and including isolator. Escalators - As detailed above. Conveyors - As detailed above.
5K - Protective Installations - Sprinkler Installations - The complete sprinkler installation and CO_2 extinguishing system including tanks control mechanism etc. Fire-fighting installations - Hose-reels, hand extinguishers, asbestos blankets, water and sand buckets, foam inlets, dry risers (and wet risers where only serving fire fighting equipment). Lightning Protection -The complete lightning protection installation from finials, conductor tapes, to and including earthing.
5L Communication Installations - Warning Installations (fire and theft). Burglar and security alarms. Fire alarms. Visual and Audio Installations - Door signals, timed signals, call signals, clocks, telephones, public address, radio, television, pneumatic message systems.
5M Special Installations - All other mechanical and/or electrical installations (separately identifiable) which have not been included elsewhere, eg chemical gases; medical gases; vacuum cleaning; window cleaning equipment and cradles; compressed air; treated water; refrigerated stores.
5N Builder's Work in Connection with Services - Builder's work in connection with mechanical and electrical services.
5O Builder's Profit and Attendance on Services - Builder's profit and attendance in connection with mechanical and electrical services.

6 EXTERNAL WORKS
6A Site Works - Site Preparation - Clearance and demolitions. Preparatory earth works to form new contours. Surface Treatments - Roads and associated footways; Vehicle parks; Paths and paved areas; Playing fields; Playgrounds; Games courts; Retaining walls; Land drainage; Landscape work. Site Enclosure and Division - Gates and entrances. Fencing, walling and hedges. Fittings and furniture - Notice boards, flag poles, seats, signs.
6B Drainage - Surface water drainage. Foul drainage. Sewage treatment.
6C External Services - Water Mains - Mains from existing supply up to external face of building. Fire Mains - Main from existing supply up to external face of building; fire hydrants. Heating Mains - Main from existing supply or heat source up to external face of building. Gas Mains - Main from existing supply up to external face of building. Electric mains - Main from existing supply up to external face of building. Site Lighting - Distribution, fittings and equipment. Other Mains and Services - Mains relating to other service installations (each shown separately). Builder's Work in Connection with External Services - Builder's work in connection with external mechanical and electrical services: eg pits, trenches, ducts etc.
6D Minor Building Work - Ancillary Buildings - Separate minor building such as sub-stations, bicycle stores, horticultural buildings and the like, inclusive of local engineering services. Alterations to existing buildings - Alterations and minor additions, shoring, repair and maintenance to existing buildings.

7 PRELIMINARIES
Priced items in Preliminaries Bill and Summary but excluding contractors' price adjustments.
This is not classed as an element but is included for allocation of costs.

8 EMPLOYER'S CONTINGENCIES
This is not classed as an element but is included for allocation of costs.

9 DESIGN FEES (on Design and Build Schemes)
9A Work Complete Before Commencement of Construction.
9B Work During Construction - These are not classed as elements but are included for allocation of costs.

BCIS Standard Elements are reproduced with the permission of the Building Cost Information Service Ltd.

Detailed Form of Cost Analysis

Blank form for submitting Detailed Cost Analysis to the BCIS, to be completed in accordance with the Principles, Instructions and Definitions of the Standard Form of Cost Analysis.

The complete Detailed Cost Analysis to be accompanied by a simple plan and elevation drawing of the scheme if available.

Cost Analysis for: _____

Submitted by: _____

Prepared by: _____

Agreement obtained for submission to BCIS: Yes / No

BCIS USE ONLY

File No: _____   Analysis No: _____

Submitted to BCIS on: _____   Drawings: _____

Entered by: _____   Date: _____

Checked by: _____   Date: _____

BCIS Ltd, 12 Great George Street, Parliament Square, London SW1P 3AD. Tel: 020 7695 1500

© RICS

HEADING	BCIS Use Only
Analysis Reference No:	

Job title: _____

Building function: _____

Type of work: _____

Date for receipt of tender: _____

Date of tender/ base month: _____

Date of acceptance: _____

Date of possession: _____

Complex contract: Yes []   No []

Main form of construction: Steel frame []   Concrete frame []
Structural brickwork []   Timber frame []

No of Storeys: Primary _____

Secondary _____

Basement _____

Gross floor area: _____ m^2

PROJECT DESCRIPTION (approximately 57 words)

_____

_____

_____

_____

_____

_____

LOCATION AND SITE   **BCIS Use Only**

Town and county: _____   Grid ref: [   ]

Local authority district: _____   Loc code: [   ]

Site conditions:

Contour:	Level/Gently Sloping/Steeply Sloping/Undulating	1 ___ ___
Ground conditions:	Good/Moderate/Bad	2 ___ ___
Excavation:	Above/Below Water Table/In Running Water	3 ___ ___
Existing state:	Green Field/Demolition/Demolition by Others/Infill	4 ___ ___
Working space:	Unrestricted/Restricted/Highly Restricted	5 ___ ___
Access:	Unrestricted/Restricted/Highly Restricted	6 ___ ___

Site description: _____

_____

_____

_____

_____

MARKET CONDITIONS (approximately 57 words)   BCIS Use Only

_____
_____   Project
_____   TPI: [   ]
_____

_____
_____
_____   Index
_____   Base: [   ]

_____

CONTRACT PARTICULARS   BCIS Use Only

Client name: _____   [   ]

Tender documents: _____

Selection of contractor: _____   [   ]

Issued:	_____	[ ]
No of tenders:	Received:	_____	[ ]

Type of contract: _____   [   ]

Cost fluctuations: _____   [   ]

Contract period: Stipulated by client: ....weeks   [   ]

                      Offered by contractor: ....weeks   [   ]

                      Agreed: ....weeks   [   ]

Tender amended: _____   [   ]
(please give details)   _____

COMPETITIVE TENDER LIST

1)_____   6)_____

2)_____   7)_____

3)_____   8)_____

4)_____   9)_____

5)_____   10)_____

BREAKDOWN OF COSTS

Contract:		Building (only if complex contract):	
Measured work	_____	Measured work	_____
Prime cost sums	_____	Prime cost sums	_____
Provisional sums	_____	Provisional sums	_____
Preliminaries	_____	Preliminaries	_____
Contingencies	_____	Contingencies	_____
Contract sum	_____	Contract sum	_____

ACCOMMODATION AND DESIGN FEATURES (approximately 95 words)

_____

_____

_____

_____

_____

_____

_____

_____

_____

_____

AREAS

		Area of external envelope:	
Basement:	_____ m^2	(ext.walls, windows & doors)	_____ m^2
Ground floor:	_____ m^2		
Upper floors:	_____ m^2	Average storey height:	
Total*:	_____ m^2	Basement:	_____._____ m
		Ground floor:	_____._____ m
		Upper floors:	_____._____ m
Usable:	_____ m^2		
Circulation:	_____ m^2	Internal cube:	_____ m^3
Ancillary:	_____ m^2	Spaces not enclosed:	_____ m^2
Internal divisions:	_____ m^2		
Total*:	_____ m^2	Number of units:	_____

*Note these totals should equal the gross floor area given above.

FUNCTION AND SHAPE

Functional Units

	Quantity	Unit/ Description (e.g. No. of Beds)	BCIS use only Code
1)	_____	_____	_____
2)	_____	_____	_____
3)	_____	_____	_____

Design Shape

	Storeys	Percentage		Storeys	Percentage
1)	_____	_____._____	4)	_____	_____._____
2)	_____	_____._____	5)	_____	_____._____
3)	_____	_____._____	6)	_____	_____._____

ELEMENT COST ANALYSIS

Element		Total Cost of Element £	Cost per m2 GFA £/m2	Element Unit		Cost Included in Another Element (state element)
				Quantity	Unit	
1	**Substructure**					
2	**Superstructure**					
2A	Frame					
2B	Upper Floors					
2C	Roof					
2D	Stairs					
2E	External Walls					
2F	Windows & External Doors					
2G	Internal Walls & Partitions					
2H	Internal Doors					
	Group Element Total					
3	**Finishes**					
3A	Wall Finishes					
3B	Floor Finishes					
3C	Ceiling Finishes					
	Group Element Total					
4	**Fittings & Furnishings**					
	Carried Forward					

Element		Total Cost of Element £	Cost per m2 GFA £/m2	Element Unit		Cost Included in Another Element (state element)
				Quantity	Unit	
	Brought Forward					
5	**Services**					
5A	Sanitary Appliances					
5B	Services Equipment					
5C	Disposal Installation					
5D	Water Installation					
5E	Heat Source					
5F	Space Heating					
5G	Ventilating System					
5H	Electrical Installation					
5I	Gas Installation					
5J	Lift and Conveyors					
5K	Protective Installation					
5L	Communications Installations					
5M	Special Installations					
5N	BWIC with Services					
5O	Profit and Attendant Services					
	Group Element Total					
	Building Sub-Total					
6	**External Work**					
6A	Site Works					
6B	Drainage					
6C	External Services					
6D	Minor Buildings					
	Group Element Total					
7	**Preliminaries**					
8	**Contingencies**					
	Contract Sum					

ELEMENTAL SPECIFICATION AND DESIGN NOTES

		BCIS Use Only	
	Table	Code	Translation
1 SUBSTRUCTURE:			Lowest floor slab
	1 7	1 Q 4	Insitu concrete bed
	1 7	4 Q 4	Basement, insitu concrete
	1 7	2 F 2	Suspended PCC
			Foundations
	1 8	4 Q 4	Insitu concrete pad
	1 8	6 Q 4	Insitu concrete strip
	1 8	7 Q 4	Insitu concrete raft
			Piling
	1 9	0	Undefined piling
	1 9	2 Q 4	Replacement insitu concrete
2A Frame:			Frame
	2 8	2 H 2	Steel column and beam
	2 8	2 Q 4	Insitu concrete col. & beam
	2 8	4 H 2	Steel portal frame
2B Upper floor:			Upper floor
	2 3	2 Q 4	Insitu concrete slab
	2 3	3 F 2	PCC composite
	2 3	3 I 1	Timber composite
2C Roof:			Roof structure
	2 7	1 I 2	Softwood flat
	2 7	2 I 2	Softwood pitched
	2 7	1 H 2	Steel flat
	2 7	2 H 2	Steel pitched
			Roof finishes
	4 7	5 F 2	Concrete/clay tiles
	4 7	4 H 2	Cladding on frame, steel
	4 7	7	Patent glazing

		BCIS Use Only		
		Table	Code	Translation
2D	Stairs:	Stairs		
		2 4	1 F 2	PCC straight
		2 4	1 Q 4	Insitu concrete straight
		2 4	1 I 2	Softwood straight
		2 4	1 H 2	Steel, straight
2E	External Walls:	External walls		
		2 1	1 G 2	Loadbearing brick
		2 1	3 G 2	Non-loadbearing brick
		2 1	F 4	Block
		2 1	4	Curtain walling
		2 1	6	Cladding
		External wall finishes		
		4 1	1 Q 4	Render/roughcast
		4 1	6	Cladding
2F	Windows and External Doors:	Windows and external doors		
		3 1	H 4	Aluminium
		3 1	I 2	Softwood
2G	Internal Walls and Partitions:	Internal walls		
		2 2	F 4	Block
		2 2	1 F 4	Loadbearing block
		2 2	2 F 4	Non-loadbearing block
		2 2	2 I 1	Non-load. timber stud
		2 2	4	Proprietary partitions
2H	Internal Doors:	Internal doors		
		3 2	3 I 2	Softwood or flush

		BCIS Use Only	
	Table	Code	Translation
3A Internal Wall Finishes:			Internal wall finishes
	4 2	1 F 7	Plasterboard
	4 2	1 R 2	Plaster
	4 2	1 G 2	Clay/Concrete tiles
	4 2	1 V	Paint only
3B Floor Finishes:			Floor finishes
	4 3	1 N 6	Plastic tiles/sheet
	4 3	1 G 2	Clay tiles
	4 3	1 J 6	Carpet
	4 3	1 I 7	Independent floor
			Secondary floors
	3 3	1	Access floor
3C Ceiling Finishes:			Ceiling finishes
	4 5	1 F 7	Plasterboard
	4 5	1 R 2	Plaster
			Suspended ceiling
	3 5	1	Undefined
	3 5	1 M 1	Mineral fibre
4 Fittings:			Fittings
	7 0	0	Normal for building type
	7 0	8 6	Special for disabled

		BCIS Use Only	
	Table	Code	Translation
5A Sanitary Appliances:			Sanitary fittings
	7 4	0	Normal for building type
5B Services Equipment:			Services installation
	6 7	1 5	Kitchen equipment
	6 7	1 8	Laundry equipment
5C Disposal Installation:			Waste disposal
	5 2	3	Soil and waste
5D Water Installation:			
5E Heat Source:			
5F Space Heating and Air Treatment			Space heating
	56	2 W 4	Central, HW Gas
	56	1W1	Local, Electrical
			Air conditioning and ventilation
	57	0	Air cond. undefined
	57	1	Heating & Cooling Central
5G Ventilating System:			Air conditioning and ventilation
	57	5	Mech. ventilation central
	57	6	Mech. ventilation local
5H Electrical Installation:			Electrical installation
	6 0	1	Power and light
	6 0	1 9	Power and light PC
	6 7	2 3	Emergency lighting/power

		BCIS Use Only	
	Table	Code	Translation
5I Gas Installation:			
5J Lift and Conveyor Installation:	Lift and conveyor		
	6 6	1 9	Lifts, PC Sum
	6 6	1	Lifts
5K Protective Installation:	Security installation		
	6 8	4	Fire fighting manual
	6 8	3	Sprinklers
	6 8	7	Lightning Protection
5L Communications Installation:	6 8	1	Burglar alarms
	6 8	2	Fire alarms
	6 8	1 1	CCTV
	6 8	1 2	Panic/attack alarms
	Special installation		
	6 7	21	Public address
	6 7	34	Data cabling
	6 7	38	Communal TV/satellite
5M Special Installations:	6 7	1	Window cleaning
	6 7	3	Medical/lab gas
	6 7	3 1	Building Management Sys.
5N BWIC with Services:			
5O Profit and Attendance on Services			
% of the remainder of the Services amount.			

	BCIS Use Only		
	Table	Code	Translation
6A Site Works:	External works (Table 90)		
	9 0	1	Site preparation
	9 0	2 F 2	Paving PCC
	9 0	2 S 5	Paving macadam
	9 0	2 W	Landscape/Plane
	9 0	4 G 2	Enclosures, brick
	9 0	4 H 2	Enclosures, steel
	9 0	4 I 1	Enclosures, timber
6B Drainage:	9 0	5	Drainage
6C External Services:	9 0	6	External services
	9 0	6 1	External lighting
6D Minor Building Works:	9 0	7	Minor buildings
	9 0	8	Inc. other buildings
	9 0	9	Work to existing building
7 Preliminaries: % of remainder of Contract Sum (excluding Contingencies)			
8 Contingencies: % of remainder of Contract Sum (excluding Preliminaries)			

CREDITS

Submitted by:

Client:

Architect:

Quantity Surveyor:

Planning Supervisor:

Engineers -   Mechanical:

                             Electrical:

                             Structural:

General Contractor:

ASSOCIATED ANALYSES

1		5	
2		6	
3		7	
4		8	

References

Andrews, J. and Derbyshire, A. (1993). *Crossing Boundaries*, CIC.

Aouad, G. *et al*. (1999). 'An IT supported new process', Betts, M. (ed.), *Strategic Management of IT in Construction*, Blackwell Science, Oxford.

Aqua Group. (1990). *Tenders and Contracts for Building*, Blackwell Science, Oxford.

Ashworth, A. (1988). *Cost Studies of Buildings*, Longman.

Banwell, H. (1964). *The Placing and Management of Contracts for Building and Civil Engineering Work*, Report of the Committee of Sir Harold Banwell, HMSO, London.

Barnes, N.M.L. (1970–71). 'The design and use of experimental bills of quantities for civil engineering contracts', PhD thesis, University of Manchester Institute of Science and Technology, Manchester.

Barrett, P. (1999). *Better Construction Briefing*, Blackwell Science, Oxford.

Baum, A., Mackmin, D. and Nunnington, N. (1995). *The Income Approach to Property Valuation*, International Thomson Business Press, London.

BCIS. (1969, reprinted 1973). *Standard Form of Cost Analysis*, The Royal Institution of Chartered Surveyors, London.

BCIS. (1984). *BCIS On-Line Users Manual*, The Royal Institution of Chartered Surveyors, London.

BCIS on line. (1998). http://www.bcis.co.uk

Benchmarking the Government Client. (1999). *Stage Two Study Document*, HMSO, London.

Bennett, J. and Jayes, S. (1995). *Trusting the Team*, Centre for Strategic Studies in Construction, University of Reading.

Bindslev, B. (1995a). 'Logical structure of classification systems', Brandon, P.S. and Betts, M. (eds) *Integrated Construction Information*, E & FN Spon, London.

Bindslev, B. (1995b). *Paradigma*, CBC systems.

Bowen, P. (1995). *Communication-based Analysis of the Theory of Price Planning and Control*, RICS Research Foundation.

British Property Federation. (1983). *The BPF System for Building Design and Construction*, British Property Federation.

Brook, M. (1998). *Estimating and Tendering for Construction Work*, second edn, Butterworth Heinemann, Oxford.

The Building Centre Trust. (2001). *Effective integration of IT in Construction*, London.

CIB Working Commission W58-SfB Development Group. (1973). *The SfB System*, CIB Report no. 22, Council for Building Research, Studies and Documentation, Rotterdam.

CIB W74 Information Co-ordination for the Building Process. (1986). *A Practice Manual on the Use of SfB: The Classification System for Project and General Information in the Building Industry*, CIB Publication 55, Council for Building Research, Studies and Documentation, Ireland.

CLASP. (1969). *Building Industry Code*, ONWARD Office.

Comptroller and Auditor General. (2001). *Modernising Construction*, H C 87 Session 2000–2001, HMSO, London.

Construction Act. (1996). *The Housing Grants, Construction and Regeneration Act*, HMSO, London.

Construction Industry Institute Australia. (1996). *Partnering: Models for Success*, Research Report 8, Construction Industry Institute Australia.

Construction News. (1997). 'New British Library: a comedy of errors', 6 March.

Construct IT. (1999). *Integrated Project Information*, Construct IT, University of Salford, Salford.

Construct IT for Business. (2000). *How to Get Started in E-business*, University of Salford, Salford.

Cooke, B. and Williams, P. (1998). *Construction Planning, Programming and Control*, Macmillan Press.

CPI. (1987a). *Co-ordinated Project Information for Building Works, a Guide with Examples*, Co-ordinating Committee for Project Information.

CPI. (1987b). *Common Arrangement of Work Sections*, Building Project Information Committee, NBS Services Ltd, Newcastle upon Tyne.

CPI. (1987c). *SMM7—Standard Method for the Measurement of Building Works: Seventh Edition*, Building Project Information Committee, NBS Services Ltd, Newcastle upon Tyne.

CPI. (1987d). *SMM7 Code of Practice*, Building Project Information Committee, NBS Services Ltd, Newcastle upon Tyne.

CPI. (1987e). *Production Drawings*, Building Project Information Committee, NBS Services Ltd, Newcastle upon Tyne.

CPI. (1987f). *Project Specification*, Building Project Information Committee, NBS Services Ltd, Newcastle upon Tyne.

CPI. (1998). *SMM7—Standard Method for the Measurement of Building Works: Seventh Edition*, Building Project Information Committee, NBS Services Ltd, Newcastle upon Tyne.

Darlow, C. (1998). *Valuation and Development Appraisal*, Estates Gazette.

Dell' Isola, A.J. (1981). *Life Cycle Costing for Design Professionals*, McGraw Hill.

Department of Employment, Trade and Resources. (1993). *Price Adjustment Formulae for Construction Categories (Series 2)*, Department of Employment, Trade and Resources.

Egan, J. (1998). *Rethinking Construction*, Report from Construction Task Force, Department of the Environment, Transport and the Regions.

ELSIE. (1988). *The Lead Consultant Expert System*, The Royal Institution of Chartered Surveyors, Quantity Surveyors Division.

Emmerson, H. (1962). *Survey of Problems Before the Construction Industry*, HMSO, London.

Enever, N. and Isaac, D. (1994). *The Valuation of Property Investments*, Estates Gazette.

Ferry, D.J., Brandon, P.S. and Ferry, J.D. (1999). *Cost Planning of Buildings: Seventh Edition*, Blackwell Science, Oxford.

Fines, B. (1974). *Building*, 25 October.

Flanagan, R. and Noman, G. (1963). *Life Cycle Costing for Construction*, Surveyors Publications, London.

Flanagan, R. and Tate, B. (1997). *Cost Control in Building Design*, Blackwell Science, Oxford.

Fletcher, L. and Moore, T. (1979). *Standard Phraseology for Bills of Quantities: Fourth Edition*, George Godwin Ltd.

Forbes, W.S. and Scoyles, E.R. (1963). *The Operational Bill, Royal Institution of Chartered Surveyors Journal*.

Fortune, C. and Lees, T. (1996). *The Relative Performance of New and Traditional Cost Models in Strategic Advice for Clients*, RICS Research Paper Series, London.

Fryer, B. (1990). *The Practice of Construction Management*, BSP Professional.

Griffiths Building Price Book. (2001). Glenigans Cost Information Service.

Gutman, R. (1988). *Architectural Practice: A Critical Review*, Princeton Architectural Press, New York.

Harris, F. and McCaffer, R. (1995) *Modern Construction Management*, Blackwell Science, London.

Hayes, R.W., Perry, R.G., Thompson, P.A. and Willmer, G. (1986). *Risk Management in Engineering Construction*, Thomas Telford, London.

Hellard, R.B. (1995). *Project Partnering: Principles and Practice*, Thomas Telford, London.

Hicks, D.T. (1992) *Activity-based Costing for Small and Mid-Sized Businesses: An Implementation Guide*, John Wiley & Sons, New York.

Higgin, G. and Jessop, M. (1965). *Communication in the Building Industry*, The Report of a Pilot Study, Tavistock Publications.

HM Treasury. (2000). *Public–Private Partnership—The Government's Approach*, HMSO, London (http://www.hm-treasury.gov.uk/docs/2000/ppp.html).

Institution of Civil Engineers. (1985). *Civil Engineering Standard Method of Measurement*, Thomas Telford, London.

Institution of Civil Engineers. (1999). *The ICE Conditions of Contract: Seventh Edition*, Thomas Telford Ltd, London.

Isaac, D. (1998) *Property Investment*, Macmillan.

Isaac, D. and Steley, T. (2000). *Property Valuation Techniques*, Macmillan.

Jaggar, D.M. (1997). *Civil Engineering Cost Analysis (CECA)*. Building Cost Information Services Ltd, London.

Kelly, J.R. and Male, S.P. (1993). *Value Management in Design and Construction—the Economic Management of Projects*, E & FN Spon, London.

Latham, M. (1994). *Constructing the Team: Joint Review of Procurement and Contractual Arrangements in the UK Construction Industry*, Department of the Environment, HMSO, London.

Laxton's Building Price Book. (2001). Architectural Press, Oxford.

Local Authority Management Services and Computer Committee. (1970). *Library of Standard Descriptions for Bills of Quantities*, LAMSAC.

Masterman, J.W.E. (1992). *An Introduction to Building Procurement Systems*, E & FN Spon, London.

McGeorge, D. and Palmer, A. (1997). *Construction Management New Directions*, Blackwell Science, Oxford.

Morton, R. and Jaggar, D.M. (1995). *Design and the Economics of Building*, E & FN Spon, London.

Mudge, A. (1971). *Value Engineering, a Systematic Approach*, McGraw-Hill Book Company, New York.

O'Reilly, J.J.N. (1987). *Better Briefing Means Better Building*, Building Research Establishment, Garston, Watford.

Property Services Agency. (1969). *Construction Planning Units*, HMSO, London.

Property Services Agency. (1990). *Schedule of Rates for Building Works*, HMSO, London.

Quantity Surveyors Division of the Royal Institution of Chartered Surveyors. (1986). *A Guide to Life Cycle Costing for Construction*, Surveyors Publications.

Raftery, J. (1994). *Risk Analysis in Project Management*, E & FN Spon.

Ray-Jones, A. and Clegg, D. (1976). *CI/SfB Construction Indexing Manual*, 1976 Revision, RIBA Publications Ltd.

RIBA. (1998). *Handbook of Architectural Practice and Management*, Vol. 2, RIBA, London.

RIBA. (2001). *Outline Plan of Work 1998*, RIBA Publications, London.

RICS and NFBTE. (1978). *SMM6 Standard Method of Measurement of Building Work*, The Royal Institution of Chartered Surveyors and the National Federation of Building Trades Employers, London.

Rowlinson, S. and McDermott, P. (1999). *Procurement Systems—A Guide to Best Practice in Construction*, E & FN Spon, London.

Royal Institution of Chartered Surveyors. (1994). *Contracts in Use: a Survey of Building Contracts in Use during 1993*, RICS, London.

Royal Institution of Chartered Surveyors. (1999). *Facilities Management and the Chartered Surveyor*, Surveyors Publications.

Royal Institution of Chartered Surveyors. (2000). *Contracts in Use: a Survey of Building Contracts in Use during 1998*, RICS, London.

Sarshar, M., Hutchinson, A. and Betts, M. (1999). 'Capability and maturity in IT and process management', Betts, M. (ed.), *Strategic Management of IT in Construction*, Blackwell Science, Oxford.

Simon, D. (1944). *The Placing and Management of Building Contracts*, Report of the Simon Committee, HMSO, London.

Skitmore, M. and Marston, V. (eds). (1999). *Cost Modelling; Foundations of Building Economics*, E & FN Spon, London.

Smith, J. and Love, P. (2000). *Building Cost Planning in Action*, Deakin University Press, Australia.

Smith, R.C. (1986). *Estimating and Tendering for Building Works*, London.

Spon's Architects and Builders Price Books. (2001). E & FN Spon, London.

Standing Committee. (1922). *Standard Method of Measurement of Building Works*, Surveyors Institution, Quantity Surveyors Association, National Federation of Building Trades Employers, Institute of Builders.

Thompson, F.M.L. (1968). *Chartered Surveyors. The Growth of a Profession*, Routledge and Kegan Paul Ltd.

Uniclass. (1997). *Uniclass Unified Classification for the Construction Industry*, RIBA Publications, London.

Vincent, S. (1995). 'Integrating different views of integration', Brandon, P.S. and Betts, M. (eds) *Integrated Construction Information*, E & FN Spon, London, pp. 53–70.

Index